T0200408

The Ontogenetic Basis
of Human Anatomy

The Ontogenetic Basis of Human Anatomy

A Biodynamic Approach to Development from Conception to Birth

Erich Blechschmidt

Revised, edited, and translated by Brian Freeman

Pacific Distributing
Murrieta, California

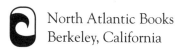
North Atlantic Books
Berkeley, California

Published by   and   Pacific Distributing
North Atlantic Books   39582 Via Temprano
Berkeley, California   Murrieta, California 92563

Cover and book design by Jan Camp
Printed in the United States of America

The Ontogenetic Basis of Human Anatomy is sponsored and published by North Atlantic Books, an educational nonprofit based in Berkeley, California, that collaborates with partners to develop cross-cultural perspectives, nurture holistic views of art, science, the humanities, and healing, and seed personal and global transformation by publishing work on the relationship of body, spirit, and nature.

North Atlantic Books' publications are distributed to the US trade and internationally by Penguin Random House Publishers Services. For further information, visit our website at www.northatlanticbooks.com.

Library of Congress Cataloging-in-Publication Data

Blechschmidt, Erich, 1904–
    [Anatomie und Ontogenese des Menschen. English]
    The ontogenetic basis of human anatomy : a biodynamic approach to
        development from conception to adulthood / by Erich
Blechschmidt ; edited and translated by Brian Freeman.
        p. ; cm.
    Includes bibliographical references and index.
    ISBN-10: 1-55643-507-X (cloth)
    ISBN-13: 978-1-55643-507-2
    1. Embryology, Human. 2. Human anatomy. 3. Developmental
        biology. 4. Human mechanics. I. Freeman, Brian, Ph.D. II. Title.
    [DNLM: 1. Anatomy. 2. Embryo—physiology. QS 4 B646a 2004a]
    QM601.B5213 2004
    612.6'4–dc22   2004006518

7 8 9 10 11 12 SHERIDAN 25 24 23 22 21

PREFACE

This book presents an overview of the changing form and structure of the human conceptus from fertilization, through embryonic and fetal stages, to the adult. The framework for interpreting the evidence is **biodynamic embryology,** where biomechanical principles are applied to the constantly changing form of the embryo and its ensembles of living cells.

Every scientific discipline requires a philosophical framework on which one can assemble the manifold data of the discipline. For the study of **human ontogeny** (development of the individual) the most commonly accepted framework is **molecular biology.** However, this type of biology mostly concerns what is happening at a molecular level and, for the purposes of comprehending human development, all too frequently imports the findings of animal experiments, findings that frequently remain untested for humans. The student of human ontogeny needs comprehensible information on the development of uniquely human cells, organs, and other supramolecular structures. In this quest, molecular biology leaves unanswered the question of *how* form arises in development and offers little other than some chemical boundary conditions. During ontogeny, the changes in the external form of the embryo reflect the positional development of new organs, and according to ongoing changes in position and form, different structures arise.

Although biomechanical embryology can be traced back to the 19th Century, it received its greatest stimulus from the dynamic interpretations of growing human embryos by the German anatomist and embryologist Erich Blechschmidt (1904–1992). This book is based on a translation of Blechschmidt's *Anatomie und Ontogenese des Menschen* (Quelle & Meyer, Heidelberg, 1978). Some figures have been revised and an Appendix has been added to explain the basis for new terms and to give an indication of the variability in the growth of the early human embryo.

Erich Blechschmidt was an anatomist and embryologist who worked for more than forty years on the problem of human form and the way that form arises in the course of ontogeny, principally during the first eight weeks or so after conception. He has provided us with more than 120 scientific papers and numerous books concerning the development of human form and function. His publications are compiled in the obituary notice written by a former pupil and colleague (Hinrichsen, 1992; see Appendix for reference). One unique aspect of Blechschmidt's research is that it concentrates on the evidence presented by the human embryo itself, in the form of a comprehensive collection of about 200,000 serial sections of embryos of different ages and sixty-four enlarged total reconstructions at the University of Göttingen (The Blechschmidt Collection and Museum). The sustained investigation of this wealth of material led to a totally new way of looking at early human development, which compels us to re-think older interpretations based mainly on phylogenetic or molecular biological studies. It is possible to see how adult functions arise naturally and consistently from the embryo's earlier growth functions. The present book is a condensation of many of these findings and a presentation of the new viewpoint.

Chapter 1 establishes the book's perspective and philosophical argument for choosing the biomechanical and biodynamical approach and introduces the ideas of growth movements, growth forces, and the **metabolic field.** Chapter 2 reviews the anatomical facts of early human development using a new rational nomenclature (explained in more detail in the Appendix). Chapter 3 describes the various kinds of metabolic fields and presents the biodynamical framework for the interpretation of the development of organs and so-called systems, which are treated in more detail in Chapters 4 to 7. Chapter 8 shows how this interpretation of human ontogeny can help in the fields of Human Biology, Psychology, Anthropology, and Sociology, by shedding light on the development of body build, instincts, gestures, language, mathematics, tools, and dress.

In keeping with Professor Blechschmidt's intention to make this book accessible to all who wish to learn about the development and anatomy of the human body, an attempt has been made to use everyday English and to minimize the number of technical terms. Nevertheless, whereas many German schoolchildren might readily grasp the meaning of, for example, *Unterzungenbogen* ("beneath-the-tongue-arch"), there is no corresponding common name in English and one is obliged to use the anatomical term *hyoid arch*. Similarly the German *Zwischenrippenmuskel* ("between-the-ribs-muscle") has been translated as *intercostal muscle* rather than the more vernacular but imprecise "rib muscle." Technical terms used in the text and in figure captions are defined in the Glossary or in parentheses in the text. Concerning the orientations and spatial relations of embryonic structures, these can often be more easily understood if one imagines oneself in the embryo's position (e.g., Fig. 2.41).

I would like to thank the Alexander von Humboldt Foundation and Dr. T. Blechschmidt for their support and counsel respectively while I was preparing this book.

—Brian Freeman, Ph.D.
Anatomy, UNSW

CONTENTS

Contents

FOREWORD

My main purpose in writing this book is to provide a suitable resource for teachers of Biology, especially Human Biology, so that they can obtain an overview of human anatomy and ontogeny. In order to define the particular point of view presented here, a view that must be considered in any contemporary study of the development of human anatomy with respect to somatic anthropology, the quantity of research evidence was less important than its quality.

This book offers, for the first time, a self-consistent presentation of human development. The presentation avoids the hitherto common-place use of specialized findings from the field of Zoology to describe *human* development. Using this book, any reader approaching human development from the discipline of general biology can quickly identify the new data and interpretations.

All illustrations are drawn from original preparations and in spite of their clarity, should not be considered arbitrary schematic diagrams. In this way, as much as possible, the object itself is given a "hearing" rather than the author. The object itself, that is, the human embryo, thereby offers the reader a reality to be admired and to be further explored, as has always been the goal of biological teaching.

I am indebted to Mr. Henn for preparing the illustrations. I would also like to thank the publishers who, with long experience in the preparation of textbooks, handled and supervised the publication of this book so carefully.

I hope that this monograph on the ontogeny and anatomy of the human being will not only fulfill the expectations of the publishers, but also the interests and needs of teachers.

—Prof. Dr. med. E. Blechschmidt, em.

August 1978

Chapter 1

General Introduction

Historical and Philosophical Background

Human nature is still a puzzle. This puzzle concerns not only the metaphysical aspects of an individual but also the somatic. The study of the form and the forming of the body is the most fundamental problem of anatomy: it applies not only to the adult, but also to the earliest stages of human **ontogeny.**[1] These days the problem has become even more acute. The growing tendency to "molecularize" the human being and to study human function in increasingly isolated "specializations" offers students little hope of "re-synthesizing" the whole from the parts. The manifold, disparate facts that chance to remain in one's memory after years of study frequently confuse rather than cohere. In this book, we will try to attain a clearer understanding of how the body is built and how its form and structure are developed and maintained by the metabolism of living cells.

Even in Antiquity, human beings were trying to comprehend the body. However, in these times the appropriate research techniques for investigating the body were quite lacking. Thus it was only in the middle of the 16th Century that Europeans began to show a scientific interest in the form of the human body and to attempt to make exact pictorial representations of it. Leonardo da Vinci's "anatomical studies," of which there are 360 drawings in the Royal Collection at Windsor Castle, are world-famous. At this time, people began to trust more their observations than the words written by earlier "authorities," such as the so-called Schools of Hippocrates and Galen. People now began to test, by dissection, ancient concepts about the structure of the human body.

Andreas Vesalius[2] (1514–1564) noted that many of Galen's anatomical descriptions were based on the study of animals and so were very confused and erroneous. Vesalius discovered that the human corpse could yield information about the construction of the living body and he was the first who, in the form of an atlas titled De humani corporis fabrica (1543), presented a description of the human body based on systematic anatomical dissection. To this day, the Vesalian anatomy of the 16th Century is the basis of the scientific practice of medicine.

Since then the human body has been investigated in increasingly finer detail. In the 19th Century, the investigator's reservoir of anatomical specimens prepared with scalpel and forceps (macroscopic or gross anatomy) was supplemented with findings based on the technique of light microscopy. Light microscopy permitted an approximately thousand-fold enlargement of the object. Today with the aid of the electron microscope, more than 200,000-fold magnifications are possible; however, only very small areas of a specimen can be examined. With the help of greater and greater magnification, we have become more and more specialized in our investigation and further and further removed from the goal of obtaining a coherent, holistic understanding of the body.

As the individual adult human arises from a single fertilized egg, the study of this development, namely **embryology**, offers the best path for a full understanding of the integration of the body. From this study, it will be seen that concepts such as body "systems" (e.g., the cardiovascular system, the nervous system, etc.) are quite artificial; their only use is for the convenience of dividing the subject matter into sections or chapters. Body systems do not exist in reality—it is always impossible to define where one system ends and the next starts. The body functions as a whole and it is only as a whole that we should attempt to comprehend it.

In the quest for knowledge, a single question has often generated many starting points, like the separate mines dug into a vast unexplored mountain. We will therefore begin our study of human development with the age-old question: How does it happen that the human body, as we know it, is constructed in its particular way and not in some other way? How does the body normally function "correctly"?

Historically, various viewpoints have been put forward as offering solutions to this question. The following interpretations of these viewpoints are offered for discussion and debate.

Functionalism: The Concept of Functional Differentiation

This viewpoint holds that the human body is built for a purpose (i.e., functionally) in the sense that the events of ontogeny should be comprehensible according to the subsequent functions of the body. As it is known, for instance, that a hen's egg invariably develops into a chick and never into a fish, one can speak of the ontogenetic process as having a *direction* by which, as it were, a design is realized. According to this viewpoint, the process of ontogeny is similar to a goal-oriented or **teleological** movement.

However, goal-oriented movements that pursue a specified purpose are volitional acts. The notion that embryological development has a purpose, in the sense of a goal, therefore presupposes the existence of a consciously acting being. This is a notion that is grounded in theology where the **teleonomy** of the universe is under consideration. However, in natural science, we should confine ourselves to considering only those processes that are biologically comprehensible. In this particular case, we should focus attention on the nature and manner of development. For the purposes of acquiring knowledge about the anatomical development of an individual, the notion of **functionalism** or expediency is irrelevant.

Whoever argues that the human hand, the claw of a cat, or the scales of fish are functionally driven differentiations is merely interpreting the structures and is not describing them scientifically. An example may clarify this argument. In one of the Grimm fairy tales, when Little Red Riding Hood sees the Wolf in her grandmother's bed, she asks, "Why do you have such big eyes?" The Wolf replies, "All the better to see you with." Little Red Riding Hood asks further, "Why do you have such big ears?" The Wolf replies, "All the better to hear you with," and finally to the question, "Why do you have such a big mouth?" the Wolf's answer is, "So I can eat you all up!" There is almost no finer example of teleological thinking than the fairy tale of Little Red Riding Hood and the Wolf.

Exactly the same kind of teleological thinking permeates contemporary morphology, especially in the field of so-called *functional anatomy*. Here human structure is interpreted according to its significance for bodily functions, that is, from the viewpoint of functionality. The claim that such-and-such a muscle flexes the elbow happens to be merely one statement from the many that could be made to describe the particular muscle, and is certainly not a statement that sheds any light on how this muscle developed to become a flexor in an embryo. If one does not carefully distinguish subjective experience from objective observation, then one will tend to draw conclusions about purposes and aims, even in a supposedly objective context, from those individual intentions that happen to have emerged into one's consciousness. As human nature this is understandable, but is not an objective approach.

The Wolf's reply to Little Red Riding Hood's question "Why do you have . . ." should have been "I have eyes and ears because they developed in me." The ability to see is not the result of a design coming to fruition but is a consequence of the embryonic development of an egg. If we say colloquially "the eye is to see with," then we are making a functional claim about the eye, roughly in the sense of "use your eyes." However, we are not providing any scientific explanation for the existence of the eye. A joint in the elbow can certainly be more useful for the execution of many tasks than a stiff arm. Nevertheless, this experience does not provide the reason for the fact that an elbow joint arises during the embryonic period. There is no doubt that many organs are indispensable for the preservation of health. However, the realization of the significance of these organs for life does not explain their origins, still less the interplay of forces involved in their differentiation.

Today we know that the later functions, or **performances**, of any organ, even those of the sense organs, are initiated through **growth functions**. Growth functions are the early activities already occurring at the time that an organ is forming. It is only on the basis of these early growth functions that the specific functions of an organ are gradually "trained." Corresponding to this, the following statement is proposed as a matter of principle: *growth functions precede all higher functions*. The achievements of the embryo are always the precursors of all subsequent accomplishments. As such the former growth functions constitute the natural plan for all adult functions. There is no cell, no tissue, and no organ that does not already function during its very development. *The fundamental functions are growth functions*. They are described in more detail in the following chapters of this book as we investigate the development of individual cells and organs. Growth functions are always dependent on the extent of the particular stage of growth and are therefore different at different ages. Thus, in the course of differentiation, each performance changes as much as the form. Ontogeny and the development of function during ontogeny are the prerequisites for the construction of the human body and all the achievements of the adult. However, ontogeny does not represent "purposeful" differentiation.

The Concept of Phylogeny and Differentiation

Prior to the 1960s when the early developmental stages of the human were still unknown and suitable methods for demonstrating early formative processes were not available, attempts were made to envisage early human development by studying the early embryos of animals. However, the techniques of histological **fixation** and staining that were available did not always permit the production of preparations that were adequate for morphological investigation.

For example, in 1859 Ernst Haeckel proposed his so-called *Biogenetic Law* according to which the human, during its ontogenetic (individual) development, is supposed to repeat or recapitulate in an abbreviated version the whole process of its phylogenetic (evolutionary) development. According to Haeckel's hypothesis, the human conceptus passed initially through early, non-human stages and then, after the development of a supposedly generalized mammalian **archetype**, finally exhibited those particular differentiations that are characteristically human.

Now no historian would ever dare to explain an event as a repetition of previous happenings. Yet in biology, one was now supposed to be able to interpret developmental processes on the basis that they were a repetition of earlier events in evolutionary history. In order to support his claim, Haeckel committed various scientific frauds (e.g., passing off an incomplete preparation of a dog embryo as a human embryo, forged diagrams, etc.). Although Haeckel later publicly admitted his forgeries, the glib expression of his hypothesis—"that ontogeny recapitulates phylogeny"—had become entrenched as scientific dogma. Even today the "law" frequently appears in popular biology textbooks without any acknowledgment of the fraud corrupting its foundation.

In any case, after the 1960s it became possible to test methodically Haeckel's ideas for the first time by applying microscopical analysis to a consistently prepared series of young human embryos. It came as a great surprise to find that the law was false. No recapitulations could be found. It is now apparent that, with the inadequately fixed or even unfixed preparations of earlier times, it was simply impossible either to reject the biogenetic law or to substantiate it. Today, due mainly to the reconstructions of early human embryos carried out in the University of Göttingen, we realize that this law is one of the greatest errors in Biology. Furthermore, we realize that Haeckel's concepts were false and that all attempts to save anything of his putative law are futile. One is not even able to claim, by paraphrasing it, that the biogenetic law is just an "approximate rule," or is only applicable in broad terms, or in a modified form.

If a contemporary chemist wishes to demonstrate conclusively that carbon combines in compounds according to its valency of four, the chemist does not argue that tetravalent carbon compounds already existed in antiquity and contemporary carbon compounds are a recapitulation of the ancestral forms. Similarly, the process of the rusting of iron cannot be explained nor understood by claiming that iron used to rust in antiquity, and that today's rusting process is a recapitulation of an earlier process. An argument such as this might indicate the frequency of rust formation but it does not explain the nature of the event. Rust always arises from specific material precursors under conditions that can be defined in more detail only by physics and chemistry, regardless of the historical significance that rust formation may once have had.

Here it is important to avoid half-truths in our concepts. One should not think that the "biogenetic law" applies today in a different form to that originally proposed by Haeckel; it simply does not apply at all. The only thing we need to know about the "biogenetic law" today is that it is false.

As mentioned above, the concept of phylogenetic recapitulation failed as soon as early human embryos were accurately investigated. Today it is known that the early stages of human development are strikingly different from the early developmental stages of all other species. Even the unfertilized eggs of different mammals are all different. The issue of why a baboon always arises from a baboon egg and a human from a human egg is clearly resolved: the egg of a baboon is already the **anlage** of a baboon and because, during its development, only its **phenotype** or manifestation changes. Ontogeny is **phenogenesis**. We talk of human development not because a jumble of cells, which is perhaps initially atypical, gradually turns more and more into a human, but rather because the human being develops from a uniquely human cell. There is no stage in human development prior to which one could claim that a being exists with not-yet-human individuality. On the basis of anatomical studies, we know today that no developmental phase exists that constitutes a transition from the not-yet-human to the human.

The argument that is sometimes advanced against this view is that one can identify processes occurring in the human embryo that are similar to processes occurring in animal embryos. However, it is obvious that processes must be occurring in every phase of human ontogeny with particular characteristics that might be similar to processes occurring elsewhere in animate, and even in inanimate, nature. Particular characteristics such as weight, volume, water-content, and peptide sequences can be the same in different species. Changes in these characteristics may also be similar. However, in the context of the whole organism, these characteristics and their changes always have a species- and an individual-specific meaning. Taken as a whole, every organism has an incomparable individuality.

In short, a fertilized human egg (conceptus) is already a human being. We appreciate that each egg contains species-specific **chromosomes** and further, we know that during ontogeny there are no mutations that would have the effect, say, of changing one species into another. It is a fact that the genetic material of an animal normally remains the same during the entire life of the individual. Without exception, a specific egg that is undisturbed during its subsequent development always gives rise to a specific organism. The following important statement is valid: that which changes during development is only the phenotype but not the essence.

According to the universal experience that one fundamental error always engenders further mistakes, belief in the accuracy of the "bio-genetic law" has led to many false interpretations and conclusions, such as the notion of so-called rudimentary or atavistic organs. So long as organs were not interpreted in the context of the individual develop-ment of the whole body, some appeared to be relics or atavisms. In real-ity, however, there are no atavistic organs. In ontogeny no organ can claim the status of a protected memorial like that of a ruin. Rather, all organs that are investigated turn out to be functioning in every phase of their development. Without exception, each organ that is develop-ing is contributing to the shaping of the whole body, even if that organ happens to undergo subsequent regression.

In analogy to Haeckel, Konrad Lorenz (1903–1989) tried to postu-late a psychogenetic principle whereby human behavior was supposed to be interpretable as a recapitulation of animal behavior. As a conse-quence of the wide acceptance of Haeckel's law, the misunderstanding easily arose that human behavior could be accounted for directly by the behaviors of our presumed animal ancestors. This belief led persuasively to the interpretation that the human being is "nothing other" than a more highly developed animal; and this led, in turn, to the view that the human can be manipulated arbitrarily and trained. Certainly there are comparable behavior patterns in different animals. However, all behaviors, be they animal or human, are always only partial processes of an individually specific whole; and this whole is the only entity with which the behaviors are directly connected. It will be shown in the following chapters that the human being displays its own unique behavior throughout ontogeny.

Having put Haeckel's law to rest, we can also dispose of the "phylogenetic account" of human ontogeny. Instead we can focus our attention on the emerging question of *how* it happens that similar organs, and similar functions, and similar behaviors can be found in humans and animals alike, despite the fact that each living creature has an inimitable individuality. It therefore becomes essential to determine the *laws* according to which differentiations occur, both in humans as well as in animals.

The Relation of Embryology to Evolution

Before considering the laws that apply to differentiations in the human embryo, it is important to clarify the distinction between embryology and **evolution**. In purely linguistic terms, evolution means development. However, evolution should be understood only in the sense of a *history*. In evolution, we are thinking of the history of the earth and of its living forms. Thus, evolution is always interpreted as a progression from lesser to higher forms.

One of the best-known philosophers who pursued evolutionary ideas was Hegel (1770–1831). Hegel's views were adopted not only by Haeckel but above all by Marx. Thanks partly to Marx, the notion of "evolution" gained ideological significance and became a fundamental doctrine of cultural and sociopolitical philosophy in the 19th Century. The sensation caused by Darwin's book *The Origin of the Species* shows how popular these ideas were at that time: on the first day of its appearance in 1859, the entire edition of about 1,200 copies sold out. However, the concept of random mutation as a sufficient principle for the process of evolution is not unambiguous. As mutations are processes occurring in the cell nucleus, the concept of evolution suggested that the origin of the species, and of life itself, was essentially an accidental process and something that took place from within the cell. According to the neo-Darwinian view, every process of differentiation should be comprehensible in principle from purely genetic rules. This idea is a frequently advocated tenet that lies at the core of certain contemporary philosophies and ideologies.

In reality, the presumption of a relationship between evolution and ontogeny (i.e., between the history of development and individual development) warrants closer scrutiny. To speak legitimately about individual development requires proof of a principle of conservation because, strictly speaking, individual development or ontogeny is understood to be only that type of development which, despite external changes, remains unchanged in essence. It is only when a living being (individual) remains the same individual during the entire duration of its existence that one can speak of a genuine development. Now if phylogeny is to be construed as a "development," it is essential to identify what is preserved during the evolutionary history of the earth when—as has been stressed since Darwin's time—new species of plants and animals have been continually appearing. If development demands a principle of conservation, then it is difficult to envisage phylogeny as being a development. For if phylogeny, and not just ontogeny, is supposed to be a development then it is necessary to prove that some essential quality of this same phylogeny also endures unchanged (i.e., remains constant). In other words, if there is to be a development of species, then something immutable should also continue to exist along with the increasing number of species. However, at the present time, nothing certain is known about any kind of principle of conservation in phylogeny.

Jacques Monod (1910–1976) considered that pure chance, in the sense of a concatenation of events that had nothing to do with each other, was the creator of evolutionary change. Monod assumed that evolution depended exclusively on molecular processes occurring with statistical probability. According to Monod, all events construed as the result of chance were to be regarded as meaningless. The objection that has been raised against Monod's view is that chance denies all personal freedoms of the individual and, in principle, rejects the possibility that other representations of the world can co-exist with a scientific view.

According to Monod, notions such as human originality or metaphysical reality had to be rejected.

Investigations of human ontogeny have shown conclusively that the notion of evolution is totally unnecessary for an understanding of individual development and anatomy. If we compare propagation on the one hand with differentiation on the other, we observe that propagation (without which the constant recurrence of development *de novo* would not be possible) is a process that begins on the inside of the cell in the nucleus and is directed to the outside. On the other hand, differentiations begin on the outside of the cell and set to work here first before gradually penetrating toward the nucleus. As we shall see, this principle applies to the whole living body, as well as to each of its cells—we can designate it as the principle of *outside–inside differentiation*. There are no hard facts to support the concept (frequently connected with the idea of evolution) that development in ontogeny is from an inferior to a superior state. An **ovum** has an extremely high capacity (potential) for original, creative development and this capacity decreases continuously with its ontogeny and therefore with growth and aging. The mature organism, although developed, is not a "higher" entity than the egg from which it grew.

It is true that findings from **paleontology** can furnish data on phylogenetic history and so elucidate an historical authenticity of a species. Yet by means of paleontological studies it has not been possible to give either an account, or even a compelling interpretation, of the structure of the human body with its specific behavior patterns. Rather, the difference between evolution and ontogeny is so great that the findings of human ontogeny had to be worked out independently of all paleontological research. It is only now, after the foundations for our understanding of human phenogenesis have been prepared, that we can commence rational comparative investigations of human and animal development.

The above argument may be clarified by an example from physics. Studying the history of lighting devices—from tapers to oil lamps to electric globes—does not solve either the problem of the properties of an electric globe or the problem of the nature and laws of light itself. Whoever wishes to know about these subjects must attempt, with physical methods, to obtain a lucid understanding of the laws of optical phenomena.

Another example: An historian may perhaps present a developmental series of vehicles extending from, say, the horse-drawn cart to a modern racing car. However, such a series sheds no light at all on the morphological and physical properties of a contemporary motorcar, or on the rules for its manufacture. Furthermore, no one would build a motorcar by first building a handcart and then gradually and progressively changing this cart into a modern type of car. No one would even contemplate recapitulating the history of motor vehicle construction, but would use immediately the appropriate processes for the construction of a car. Any and every recapitulation of outmoded methods and models would carry with it the risk of bankruptcy for the factory!

There is no doubt that without the history of the earth, our presence on it would be inconceivable. Nevertheless, the human body, for all its complexity and regularity, cannot be described in detail using concepts from evolution, but can only be understood as a consequence of its own ontogenetic development. Indeed, the history of development on the earth could be interpreted as the sum of all ontogenies. However, of these, only the ontogenies of contemporary living species can be investigated and known with any precision.

One cannot draw conclusions about processes occurring in ontogeny by studying phylogeny. What we call ontogeny cannot be compared to, say, the construction of a city by renovating parts of an earlier city, but rather to the actual re-establishment of a city that does not take over one single pre-existing building.

Developmental Physiology and the Concept of "Induction"

Wilhelm Roux (1850–1924) is considered to be the founder of developmental physiology. By means of animal experiments, developmental physiologists have attempted to analyze the processes of development. In so doing they have obtained many results that are important with respect to the possible factors causing malformations. However, the experiments do not provide a logically consistent description of the course of normal human phenogenesis, nor do they yield a comprehensible concept of the human body.

The biologist Hans Driesch (1867–1941) carried out an experiment in developmental physiology that has become quite famous. Driesch took the fertilized eggs of sea urchins and, by shaking them at the two-cell stage, separated the two cells (known as **blastomeres**). He then allowed each of the artificially separated blastomeres to pursue its own development. The result was the production of two complete larvae of about half the normal size. Driesch concluded from these experiments that each cell had the capacity to develop by itself into the whole organism. Yet there was certainly nothing said about *how* this capacity for development resulted in the actual development—no claim at all was made concerning the "how." In general, it was simply left unanswered as to what was meant by the capacity to develop on the one hand and the process of development as phenogenesis, on the other. If a capacity for development means a capacity that is only realized under special circumstances, then it is essential to explain which particular circumstances are created (e.g., experimentally) so that the actual development takes place.

In this respect, we should mention another famous experiment by the zoologist Hans Spemann (1869–1941). Spemann experimented on amphibian eyes in order to understand the causal origin of the so-called eyecup. In a manner similar to what is observed in many other species, the amphibian eyecup develops from a blind-ended vesicular **evagination** of part of the wall of the brain. In contrast, the lens of the amphibian eye forms later and develops directly from the overlying **ectoderm** (this also occurs in the development of the human eye, described in Chapter 5 and illustrated in Fig. 5.6). Spemann now carried out the following experiments. He destroyed the eyecup of a grass-frog tadpole before a lens could develop, thereby preventing lens formation. His conclusion was that the lens develops in dependent-differentiation (i.e., dependent upon the eyecup): if the eyecup is missing, no lens develops. Later Spemann repeated the experiment on a related cold-blooded species, namely on the tadpoles of the water-frog. However, in this case a lens did develop. Spemann concluded that in this animal the lens develops independently of the eyecup—an example of independent- or self-differentiation.

Further experiments revealed similar results. If a few cells were transplanted from the body region of a salamander where normally a tail would grow, to the region where normally a leg would grow, then the tail cells would develop into a leg. The resulting conclusion was: dependent-differentiation. However, if the tail cells were transplanted after they had reached a later stage of growth, then a tail was found to develop; this event was described as independent-differentiation. Further experiments based on Spemann's investigations led to the following generalized views. Young, still undifferentiated cells can be stimulated through contact with their environment to differentiate according to their location and adapted to the surrounding structures. On the other hand, more fully developed cells are less adaptable and are therefore regarded as "determined." On the basis of his experiments Spemann concluded that, especially in the early stages of development when a particular differentiation appears to be dependent, there exists an organizational center or an *organizer*, which brings about the differentiation. Material influences were supposed to emanate from each of the anlagen or primordia, which had already developed to a particular degree, to induce further development in the organism. The results were ascribed to the effects of the organizer. The issue of the kinetics and the dynamics of the developmental movements therefore remained unconsidered. The effective principle of the organizer was considered to be in the form of chemicals called inductors or morphogens, capable of bringing about a particular **induction**.

However, to this day, the natural existence of "inductors" has never been confirmed: no specific chemical substance that causes a particular differentiation has ever been identified. The organizer remains a *deus ex machina*.

On the contrary, exactly the opposite proposition has been proven repeatedly, namely that specific so-called inductions can be caused by many different organic and inorganic (often strikingly non-specific) substances. In addition, it has been shown repeatedly that simple mechanical interference to the ovum can lead to specific changes. For example, the insertion of a tiny rod of whalebone (i.e., the horny substance of a whale's barb) or a sliver of paraffin wax into an embryo may achieve a particular differentiation. In rabbits, the development of the ovum to a fully grown fertile female animal can be accomplished by pricking the unfertilized egg with a steel needle then re-implanting the egg in the uterus. This is the phenomenon of artificial **parthenogenesis,** which also occurs with some amphibians where development of the egg may ensue after suddenly changing the concentration of salt in the surrounding water.[3]

It is now certain that specific inductors, or immediately effective morphogens, do not exist. We therefore still have to confront the following questions: What are the conditions that enable the processes of differentiation to take place? What are the general rules that govern these processes?

Molecular Biology and the Concept of "Genetic Information" in Embryology

In the course of his famous crossbreeding experiments, Gregor Mendel (1822–1884) discovered the laws of inheritance that are now named after him. At that time the nucleus of the cell was still an unknown. Today we not only know about cell nuclei and chromosomes but we also have precise concepts about their structure. The biochemist Erwin Chargaff (1905–2002) was the first person to propose a model for the chemical structure of deoxyribonucleic acid (DNA), which is the genetic material present mostly in the cell nucleus. DNA is a giant molecule that is conceived of as a thread, which can be split lengthwise and then can be duplicated by adding complementary bases to the split halves. As is generally accepted today, the transfer of genetic material from one generation to the next and thus the availability of this material for the daughter cells depends on this property of DNA-splitting and its invariant reduplication. It has become customary to describe

DNA as having a "genetic code" whereby the arrangement of the bases of the nucleic acids is of direct significance for the construction of cellular proteins.

And yet even today, frequently used concepts such as "instruction" by the **genes** or the "recall" of genetic information are still a source of unsolved difficulties. It is accepted that it is not the genetic substance itself, but rather the manner in which the genetic substance is decoded or read-out that determines the role of the genes in differentiation. For example, the entire linearly ordered genetic code could be read serially or partially. In humans, on account of the intermingling of the chromosome threads, the concept of partial decoding is of great significance. In very simple terms, geneticists envisage that these processes depend on regulator substances that either free or block the reading of the code.

In spite of the elegance of this concept, the problem of *how* the chemistry of the genes is related to the forming of the body remains as completely unsolved as before. Nevertheless, one thing is clear, namely that the "read-out" always takes place only with the participation of the *extragenetic material* that is already present both in the nucleus and in the **cytoplasm**.

The DNA might be regarded as a cookbook that the body requires for its metabolism. However, it is unclear who the cook is in the kitchen. And it is particularly unclear how the body can decide, from the wealth of available recipes, which are the "correct" genes to use. Just because genes happen to exist, it does not follow logically that the course of the organism's development and all its differentiations arise from the genes in a sequential manner. All the knowledge in the universe of gene molecules cannot solve the question of *how* cells differentiate.

Any theory that alleges that the structure of the human body is based primarily on the information contained in the genes is quite dubious. Why? Well, consider the following: in any organism, almost all cells are equipped with identical genes. Now these genes, which are the same in almost every cell of the human body, would have to know by themselves on the basis of their "information" how, where, and what differentiations should occur, in each split-second, in every part of the organism, during the whole course of the organism's life. The notion is ludicrous. Consider further the proposition that the genes themselves are parts of large molecules, which are cellular constituents with changing forms and functions. By the same "information" argument, these structures would have to be controlled by a gene, and this gene by another gene, and so on. The *reductio ad absurdum* indicates that our simplifications based on "genetic information" lead to a false concept of the cell. Such simplifications totally ignore the integrated nature of the cell, and in particular, they ignore the fact that the genes have to work in harmony with all the other molecular components of the cell and its environment.

There is no doubt that the genes are of great importance for heredity and that they therefore play a role in differentiation. However, even in the young **germ**, neither the chromosomes nor the genes are dynamically active. Genes are not the motors of development and therefore do not themselves evoke the characteristics of the differentiated organism. Genes do not work, even indirectly, via the **enzymes** they help to make. To believe in this last view only shifts the whole problem of development and differentiation from the chemistry of the genes to the chemistry of the enzymes. Even the most important enzymes are only aids to the process of differentiation, an event that has many other prerequisites apart from the genetic information itself. The cell is much more than the genes. The cell only functions effectively as a whole (nucleus plus cytoplasm and organelles plus boundary membrane) and it is only in the framework of this whole that the genes have any meaning.

There is no direct connection between the genes in a nucleus and the form of the whole cell. Just as it is unlikely that a sandbank, by itself, can cause ripples at low tide on the basis of its chemical nature alone, but rather requires the strength of the wind and the surge of the waves, so it is equally unlikely that formative materials, rather than formative forces, are the direct motors of phenogenesis. It is the task of *kinetic morphology* to provide an understanding of these motors. In any phase of development, changes in form and structure must result from the complex movements of particles of a molecular and submolecular nature. At all times, such movements, which are the manifestations of physical forces, are the *direct* causes for those changes in position, form, and structure that lead to differentiations. As we know, every chemical reaction, every change in metabolism, also has (bio)physical components. And it is precisely upon such components that phenogenesis depends. As we are concerned with living organisms, the most fundamental forces associated with these components can be described as **biodynamic**. Thus the form of the organism differentiates *directly* under biodynamic forces, not chemical–genetic information.

What then is the contribution of the genes? Today we realize that the genetic material attains its significance in the context of the individually specific living *metabolism* already occurring in the ovum. The genes do not contain a pattern for subsequent differentiations. The belief that the genes represent a blueprint is now out of date. The totality of the genes, as the **genome**, does not constitute a **homunculus** that resides, say, partly in the nucleus of the ovum and partly in the sperm.[4] The genes do not cause differentiations. Rather, they are the chemical constants of metabolism and, as such, are especially stable components in a cell. It is certain that the genetic material serves to preserve the individuality of the organism, whereas vice versa the *extragenetic material*, particularly the cytoplasm, elicits the changes in appearance that are observed during development. In the course of differentiation (phenogenesis) in a multicellular organism, it is probable that the different effects of the extragenetic material on the genome are much more significant than influences in the reverse direction (i.e., from the cell nucleus to the cytoplasm). As the genes are such especially stable structures, they are much too inert to act by themselves. Rather they respond to the stimuli coming from the cytoplasm (i.e., from "outside"). They do not act, so much as react. Genes are a necessary prerequisite, but not a sufficient condition, for the process of differentiation.

This presentation may appear to be unusual. However, the following example can clarify the argument. We know that the effect of a force depends on what exists at the point of its application. For instance, pressure can crack a sheet of glass or make a bell ring. Through carving, pressure can also impart an entrancingly beautiful exterior to a piece of Carrara marble. In every case, what actually arises from the action of pressure depends on the circumstances at the point of application. The same holds true for human differentiation. In the interplay of the forces of cellular metabolism, a cell nucleus with its so-called information carriers is the stable reference system for all developmental stimuli. This frame of reference establishes an important factor for the preservation of human individuality during development. Therefore, the sum total of development is much more than the genetic information alone.

Another example, this time from the inorganic world, may further clarify the argument. Iron is obtained by mining ore and can be the basic material for the manufacture of nails, iron lattices, watch springs, and many other metallic products. All these industrial products can be better understood in their diversity if the nature of their production is known. Thus if we want to know, for example, why (in the material sense) fine watch springs vary in their suitability for the production of quality watches, then we need to know the *material constants* of the different kinds of iron used in the manufacture of the springs. In biological terms, we would need to know their different "genetic properties." And yet it is clear that temperature, temper, and many other physical factors are of great significance in determining the usefulness of these springs in a watch. The intrinsic material characteristics ("constants") often become effective only in the final stages of a particular **functional development**.

In biology, it is known that chemical structures can change under different physical conditions. This means that kinetic and dynamic processes (the distributions of tension, temperature, etc.) can influence the extragenetic material, which in turn leads to a reading of the genetic code.

The "information" in the cell nucleus represents the following: *the preservation of individuality with the help of an individually specific metabolism*. In the most general sense, genetic information denotes the capacity to be human. In order to develop the human phenotype with the aid of this "information," the extragenetic material, where the metabolic processes are actuated, requires the use of the genetic material.

In the interplay between non-specific stimulus and specific reaction, the macromolecular genes regulate the effects of the stimulus, which might otherwise represent a disruptive influence. These compensating processes express themselves as the different steps of development. The substrate for all these processes is probably an intracellular circulatory system of metabolic movements from the cell boundary membrane toward the nucleus and then back toward the cell membrane. This intracellular circulation can be considered to be a precondition for the functional connections between the anlagen in the interior of the cell and their external expressions in the phenotype.

On the basis of investigations in human embryology, we are able to state today that phenogenesis, for the most part, is traceable to laws that are not embodied in the genetic substance of the germ cells. The extragenetic (phenogenetic) substrate, which is based mainly in the cell cytoplasm and boundary membrane, conforms to biodynamic laws that cannot be reduced to genetic information. Nevertheless, this substrate probably influences the genes in different ways at different times and places during ontogeny.

Differentiation in Biodynamic Metabolic Fields

In the preceding sections, we have seen that many attempts to comprehend the form and structure of the human body from solitary viewpoints have not succeeded. We now enquire further: How does the embryo at a particular time in phenogenesis regularly manufacture body components such as shoulders and arms? How does it occur that the mouth is constructed as a horizontal aperture and not as a vertical one? How does the embryo perform so that its eyes are located at the front of the head and not at the back? In short, how does it happen that we are shaped ontogenetically in the form in which we know ourselves and not in some other form?

These are valid anatomical questions. We have already explained that various answers such as "to achieve a goal," or "because of an historical pattern," or "because of a genetic blueprint" do not provide any real information because they do not describe in what manner the fertilized ovum changes its phenotype step-by-step.

If one wants to obtain a conception of the human being from a scientific point of view, then one has to grasp the relationship between the *form* and the *forming* of the body. If we want to describe the body not as a state but as a process of formation, then we must follow the spatial changes in the body's form over defined periods of time. These changes are **developmental movements**. In so doing, we may initially postpone all biochemical investigations.

Studying the developmental movements and postponing biochemical investigations does not mean that we are guilty of reducing the complex organic process of formation to some isolated arcane aspect of the physics of the human germ. Once again an example may elucidate the argument. The distance covered by a mountaineer can be measured by geometric methods and the corresponding speed of the climb can be determined by physical methods (e.g., with a watch). The work performed can also be determined and, independently of this, chemical processes in metabolism, oxygen consumption, carbohydrate combustion, and enzymatic action could also be investigated. However, a temporary postponement of one or other of these investigations does not falsify the actual and completed study of the distance traveled or work performed. Moreover, just as chemistry is based on physics and builds on it, so chemical investigations in the biological field are only systematic and useful if data on spatial and physical properties are already available.

If by using **biomechanical** methods, one succeeds in deriving a comprehensive description of the formation of the body, then this does not mean that a contradiction must arise between such a description and the results of subsequent investigations in the fields of biochemistry or comparative anatomy. The living organism is always a whole, no matter how many different ways we view it or assess it. This is a very important statement that is worth rephrasing: *To account for the developmental process in its entirety requires more than just a physical or just a chemical description.* After all, purely physical or purely chemical processes simply do not exist in living organisms. Faced with life and all its manifold properties, we will simply be describing certain aspects of the biophysical and biochemical features of the living organism, and nothing more.

Today some of the **biokinetic** and **biodynamic** characteristics of differentiation are already so well known that it is possible to provide a logical description of the forming of the body as a process of differentiation. Instead of a mere list of findings, this description allows us to make a first step toward comprehending the human body. For each differentiating organ, one is able to discriminate between the development of the position of the organ (**topogenesis**), the development of its form (**morphogenesis**), and the development of its internal structure (**tectogenesis**). Changes in position are directly linked to changes in form, and these in turn lead to changes in internal structure. Development of position, form, and structure emerge collectively as forming movements or **forming functions**. As soon as we describe these movements, the previous study of static anatomy (the anatomy of forms) becomes one of kinetic anatomy (the anatomy of forming). Movements of molecular and submolecular materials, otherwise invisible, express themselves as the forming movements. We can say that developmental movements include those movements of materials that are not immediately visible (submicroscopic) because they occur at a molecular or submolecular level.

As developmental movements are occurring against resistances, and are also often overcoming these resistances, the movements themselves signify early living achievements and therefore represent work performed in a physical sense. The forming achievements are characteristics of the growing organism, performances on which depend all subsequent accomplishments of the adult. Without them, no mature organism could function.

Following an approach adopted in modern physics, one can describe biological processes as taking place in force fields by using the concept of a **metabolic field**. A **biodynamic metabolic field** is a field of force based on a locally ordered metabolism. Metabolic fields are those morphologically definable regions, at all different levels of spatial resolution, which contain spatially ordered metabolic movements. Biodynamic metabolic fields can be used to describe cells and cell ensembles (e.g., zones of loose tissue, zones of dense tissue) or whole areas of differentiation such as the lung, the liver, or the thyroid gland.

Until recently, the technical prerequisites that now enable us to investigate systematically the early stages of human development were completely lacking. In the early stages of development, human embryos are still so minute that in general they are hardly noticeable and are only observed during surgical procedures in exceptional circumstances. These young embryos are just as transparent as young ova and, on account of their high water content, appear almost structureless. To prevent rapid disintegration (**autolysis**) during investigation, the embryos have to be fixed seconds after death in specially prepared solutions, so that they can be studied subsequently in thin sections under the microscope.

Anatomical preparations of young human embryos still have an almost irreplaceable research value. The best specimens are serially sectioned, collected, and catalogued in large international collections, such as the Carnegie Collection in Washington, DC. However, to obtain a good spatial concept or even to make accurate diagrams from thousands of such thin sections requires the laborious production of three-dimensional reconstructions from the sections. Furthermore, it is only through a series of total three-dimensional reconstructions that one is able to determine the developmental movements of whole embryos. To prepare such *total reconstructions*, the microscopic serial sections had to be uniformly enlarged, section by section, to make flat representations that could then be cut out of polymer sheets. These plate-like models could then be reassembled to make a so-called total serial-section reconstruction (Fig. 1.1). The only standardized collection of such total reconstructions of human embryos is on display in the Anatomical Institute of the University of Göttingen[5] (Fig. 1.2). The height of any one individual model is 80 cm. Each plate in an individual reconstruction is 1 mm thick. The beeswax employed previously for embryo reconstructions was far too soft and sensitive to temperature change to be useful for the construction of such large models in a uniform series. With earlier techniques, one could reconstruct only single organs or parts thereof and consequently it was impossible to obtain any concept of overall development.

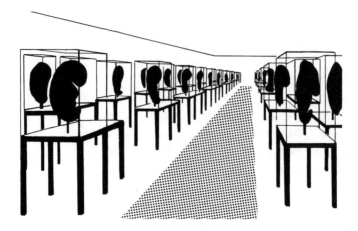

Fig. 1.1. Total reconstruction of a 10 mm long human embryo at Carnegie Stage[6] 16. The 80 cm high "model" is reconstructed from hundreds of 1 mm thick single polymer plates, each of which is a representation or model of a single histological section.

Fig. 1.2. The Blechschmidt Collection of Human Embryos in the Anatomical Institute of Göttingen University. In one large room stand 64 of the serial-section reconstructions described in the text. Total reconstructions of embryos of different ages permit spatial comparisons between single developmental stages, and this in turn permits the demonstration of developmental movements.

For the first time, with the help of the total reconstruction models, it was possible to compare the early stages of whole human embryos with one another and so determine the developmental movements leading to the formation of the human body. As well as obtaining precise knowledge on the external form of the whole embryo, it now became possible to learn about its internal structure as well. These investigations demonstrated the remarkably close interrelationship in development between external form and internal structure, that is, between morphogenesis and tectogenesis.

From the outset the total reconstructions permitted a so-called *regional comparison*, i.e., an anatomical comparison of structures from different regions of the body. For the first time in human embryology, this led to the determination of the developmental movements defined above and to the comparison of **ontological organs** (i.e., organs that arise from the one fertilized ovum). Only then was it possible to perceive the similarities between body parts and organs that were formerly considered to be so different as not to warrant comparison. Now one could recognize the common aspects of organs in the body arising as spatially different modifications of cellular ensembles. The method of regional comparison led eventually to the discovery of common *rules of differentiation*, which are described in detail in Chapter 3.

The question that will occupy us in the remainder of this book is: How does the differentiated human body arise? In this quest, we will not be asking for its evolutionary history or for its chemistry, but only for its biokinetic characteristics. We will find that a uniform pattern of cyclical and interdependent movements is of special significance for the preservation of individuality.

Chapter 2

Early Human Development

This chapter concerns early human development immediately after fertilization. Specifically, it concerns the first of the following three periods: the period of *early development* of the whole conceptus (one to three weeks after fertilization), the period of *embryonic development* within the conceptus (four to about eight weeks), and the period of *fetal development* (about three months to birth). The early and embryonic periods are described in detail in this book because they demonstrate the precision of developmental movements, even in the first few days after fertilization. It will be shown that these movements are already performances of work by the conceptus. The description will make it clear that the conceptus is always a whole, and that its differentiations represent subdivisions and modifications. As such, these differentiations are always **ontological** and so can be compared with one another. Comparison of these differentiations reveals the principles of development, thereby enabling us to determine the rules of development.

Prior to fertilization, the sperm cells in the **uterine tube** undergo a structural and chemical change called **capacitation**, whereby exposure to the fluids of the female genital tract gives spermatozoa a capacity to fertilize. This process normally takes five to six hours. At **fertilization**, the structural change in the sperm cell involves primarily its **acrosome**: by the time the capacitated spermatozoon is in the vicinity of the **ovum**, acrosomal enzymes are liberated through perforations in the acrosomal membrane. One sperm penetrates the ovum's tough **glycocalyx** capsule, known as the **zona pellucida**. Contact of the male and female germ cells results in membranous folds at the surface of the ovum; these folds envelop the head of the spermatozoon. A reactive change, known as the cortical reaction, also occurs in the peripheral cytoplasm of the ovum, just under the boundary membrane. This reaction is characterized by "boiling" movements of the cortical cytoplasm in the living ovum and the release of the contents of cortical granules; this occurs while the acrosome of the sperm is releasing its enzymes. The cortical reaction causes a change in the zona pellucida so that it now acts as a barrier to the entry of additional sperm cells. Fertilization is completed when the two nuclear masses (of ovum and spermatozoon) fuse and re-establish the standard **diploid** number of chromosomes. The fertilized ovum is called a **conceptus**.

First Week: Development of the One-Chambered Conceptus

A fertilized human ovum is about 0.15 mm in diameter, weighs approximately 0.0005 mg, and looks like a tiny drop of water. For the first three days after fertilization, the conceptus lives in the uterine tube; it is still enclosed in its delicate capsule of zona pellucida, which only breaks down after about the 3rd day (Fig. 2.1). After 40–50 hours, the conceptus becomes two-celled: the fission into two daughter cells (**blastomeres**) is evident externally by the formation of the so-called cleavage furrow and the production of a tiny volume of intercellular (extracellular) fluid that accumulates in **interstices** under the capsule. These changes signify a rearrangement at a molecular level in the conceptus: at the expense of the cytoplasm, the nucleus of the fertilized ovum grows and duplicates. Substances probably move from the cell boundary membrane to the cell nucleus and back again. These minute material movements represent an *intracellular circulation* (Figs. 2.2, 2.3). During the formation of the cleavage furrow, the conceptus does not increase in volume. It has been demonstrated that the two blastomeres exhibit different metabolism and have a different histochemical profile.

About the 3rd day, the number of cells increases rapidly, so that by the 4th day there are more than one hundred blastomeres. With the increase in the number of cells through subdivision, the total surface

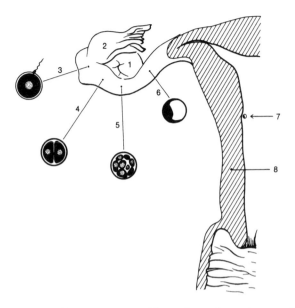

Fig. 2.1. Human uterus with uterine tube and ovary. The size of the dot at end of indicator lines is about three times the natural size of the different developmental stages of the human conceptus during the first four days after fertilization. 1) ovary, 2) ampulla of uterine tube, 3) fertilization, 4) two-cell stage, 5) blastomeric conceptus (approximately 50 cells inside zona pellucida), 6) blastocyst, 7) blastocyst at beginning of adplantation, 8) wall of uterus.

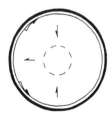

Fig. 2.2. Human conceptus within its capsule (zona pellucida, black line). Arrows schematize the pathway of an intracellular circulation of material movements whereby the cell boundary is coupled to the cell nucleus and vice versa.

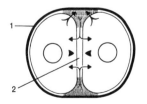

Fig. 2.3. Commencement of subdivision (fission) of conceptus into two daughter cells (blastomeres). Arrows represent metabolic movements. Stipple: secretion products of metabolism. Black line: capsule (zona pellucida).

area of the cellular ensemble and the number of nuclei also increase, as does the collective volume of the small amounts of intercellular fluid produced with each subdivision. All these increases occur at the expense of the original net volume of cytoplasm. Initially, there is no demonstrable increase in the total volume of the conceptus. Other than at sites where pockets of intercellular fluid accumulate, the newly formed cells do not separate from each other but remain enclosed in the zona pellucida, clinging to each other through the reciprocal exchange of materials (Figs. 2.4, 2.5). This metabolic exchange is a consequence of the chemical and structural disparity of daughter cells described above. It appears that these metabolic interchanges enable the cells to retain their various different forms. In this we see an important **forming function** of the very first cells.

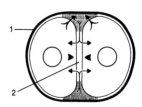

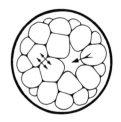

Fig. 2.4. Two-cell stage of human conceptus, still in a closed capsule. Tailed arrows indicate material movements in the sense of a reciprocal biomechanical cohesion (through metabolism) between the two blastomeres, which have arisen from subdivision of the fertilized ovum. Arrowheads represent forces due to the pressure of the cell cytoplasm. Stipple: intercellular catabolites that can be recycled into the cytoplasm. 1) capsule with underlying cell boundary membrane, 2) intercellular space.

Fig. 2.5. Transformation of blastomeric conceptus into blastocyst through eccentric pooling of fluid (a symmetrical pooling of fluid in the center of the conceptus is impossible due to the asynchrony of blastomeric division). Double arrows indicate reciprocal attraction of cells. Tailed arrow: formation of intercellular substance.

Sometimes the term "**morula**" is used to describe the cellular ensemble of this type of **blastomeric ovum**. Actually, a morula is a free-swimming, unenclosed, tight mass of fertilized sea-urchin cells. Unfortunately, to describe a human conceptus as a "morula" diverts our attention from the important biomechanical (formative) roles of the enclosing zona pellucida and the enclosed intercellular fluid pools accumulating between the blastomeres. The human embryo never exhibits the features of a sea-urchin morula, either in form or structure.

As mentioned above, there is a gradual increase in the quantity of metabolic by-products that accompanies the increase in cell number during the 3rd and 4th day. These by-products collect together eccentrically in the conceptus forming pools of fluid between the cells (**intercellular substance**, Figs. 2.4, 2.5). The pooling of this fluid within the mass of cells leads to the formation of the **one-chambered conceptus** or **blastocyst**. The eccentric location of the intercellular material is probably a consequence of the lack of symmetry in the timing of blastomere formation and the resulting variation in the thickness of the blastocyst wall: even the first subdivision produces structurally and chemically asymmetrical blastomeres. Pure symmetry in biology is virtually impossible and the increasing asymmetry signifies a polarization of the conceptus. It now possesses an *assimilation pole* and a *dissimilation pole*. At the dissimilation pole, fluid is exuded from cells that decrease further in size. With increasing accumulation of fluid in the **lumen (coelom)** of the blastocyst (**blastocoele**) the small cells become flatter (squamous).

From this we conclude that the fluid possesses an **osmotic pressure**. At the assimilation pole the cells remain larger, constituting the **blastodisc** or so-called **inner cell mass**. It is likely that these cells re-absorb substances from the fluid in the lumen of the blastocyst (i.e., from the blastocoele). Initially there is hardly any absorption of nutrients from outside the blastocyst, because an obvious increase in blastocyst volume cannot be demonstrated. This strongly suggests that, even at this stage, the changes that are taking place are more in the nature of intermixing rather than new formations (*de novo* synthesis or neogenesis). The above developments signify a differentiation into opposites, namely the formation of larger assimilating cells as opposed to smaller dissimilating cells. This phenomenon of differentiation into opposites is also a characteristic of later phases of development.

On about the 4th day, the conceptus is situated with its thick-walled assimilation pole adjacent to the mucosa of the uterus; this is known as

adplantation (Fig. 2.1). Up to the time of adplantation on the uterine mucosa, the blastocyst continues to remain about as small as the original one-celled anlage (i.e., about 0.15 mm in diameter). Its consistency is almost liquid. Around this time, the zona pellucida ruptures and the blastocyst is said to "hatch." Experiments show that hatching may result, not only from a local dissolution of the zona pellucida, but also from pulsations of the blastocyst.

When the blastocyst has attached itself to the uterine mucosa, it absorbs nourishment via its thicker wall. This nourishment is taken not only from the interior of the conceptus (i.e., from the fluid of the blastocoele), but also externally from the uterine mucosa. The blastocyst always orientates itself so that the assimilation pole is adjacent to the uterine mucosa. With the absorption of nutrients, both internally and externally, the thick region of the wall (blastodisc) expands, bulging slightly both to the inside and outside of the conceptus (Fig. 2.6). Now, for the first time, the volume of the conceptus starts to increase.

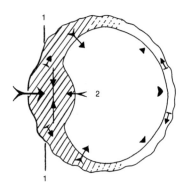

Fig. 2.6. Blastocyst at beginning of implantation in uterine wall (1) at about 4th day, diameter about 0.15 mm. Tailed arrows: uptake of nutrients into the thickest part of the blastocyst wall (the blastodisc or inner cell mass) and secretion of metabolic products into lumen of blastocyst (blastocoele). Converging double arrow: reduced growth (growth resistance) in a zone with less nourishment containing the anlage of epithelium of amnion (i.e., roof of amniotic sac). Arrows with cross-tails: growth in surface area of blastocyst. Arrowheads: fluid pressure in blastocoele (2).

Second Week: Development of the Two-Chambered Conceptus

The conceptus therefore actually sucks its way onto the uterine mucosa. As the conceptus orients itself and nestles onto the uterine surface at this assimilation pole (a process called **nidation**), it comes into direct contact with the metabolic fields of the maternal tissues. The **placenta** will form later at this interface. As is known from clinical studies, the blastocyst secretes enzymatic metabolic by-products (**catabolites**) into the maternal tissues, leading to the local destruction of maternal cells and thus the liberation of additional nutrients for the growing conceptus. With the uptake of nutrients from a virtually unlimited source, the conceptus grows explosively and actively sucks its way further and further into the maternal mucosa. Initially the most rapidly growing part of the conceptus is its outer layer, which is designated the **ectoblast**.[7] At the beginning of the 2nd week, the conceptus has almost fully **implanted** itself below the surface of the uterus.

As the blastodisc is absorbing nutrients from the outside as well as from the inside (i.e., from the blastocoele), the cells in the interior of the blastodisc receive the least nourishment. The growth of these central cells is therefore retarded relative to the growth of the cells at the outer and inner surfaces of the disc. This means that these central cells, which are adherent to their neighbors, become strained under tension as the whole blastodisc enlarges. In opposition to this tension, the central cells now develop a resistance to further stretching. The tension at their cell membranes results in a flattening of these cells. The thin layer of flattened cells that forms in this way inside the blastodisc is the **anlage of the amnion** (Fig. 2.6). The layer of neighboring cells adjacent to the amnion now tends to arch away toward the center of the blastocyst. As the curvature of this layer of cells increases, the layer gradually lifts away from the adjacent amnion to create a fluid-filled cleft, which is called the **dorsal endocyst vesicle** or **anlage of the amniotic sac**[8] (Fig. 2.7). The dorsal endocyst vesicle may be said to contain the dorsal **blastemal** fluid. By analogy, the fluid of the original blastocyst is now called the ventral blastemal fluid. The former is the

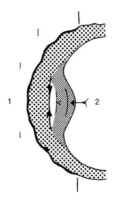

Fig. 2.7. Formation of amniotic sac in the consolidated region of the blastocyst. Coarse stipple: anlage of chorion (ectoblast = trophoblast). Fine stipple: anlage of the ectoderm and endoderm. Converging double arrow: growth resistance (restraining function) of the developing amnion. Tailed arrow: nutrient-uptake on inner side of thickened part of blastocyst wall. I) uterine wall, 2) blastocoele.

anlage of the **amniotic fluid** in the amniotic sac and the latter is the **anlage of the yolk sac fluid** in the **yolk sac**. In summary, in the 2nd week of development, two chambers or sacs of fluid (ventral blastemal fluid and dorsal blastemal fluid) arise in the conceptus. The terms "dorsal" and "ventral" describe the location of the fluid with respect to the orientation of the embryo that will form between them.

What we observe is the manifestation of spatially ordered metabolic movements that take place in the metabolic fields of the early conceptus. To survey these fields, it is sufficient to indicate some of the main directions of metabolic movements by using arrows (e.g., see Fig. 2.6ff.).

The cells between the dorsal and ventral blastemal fluids constitute the so-called **endocyst disc** (i.e., a disc lying within the system of vesicles). This disc is the **anlage of the embryo**. In the period of early development, the endocyst disc is double-layered (bilaminar). The floor of the dorsal endocyst vesicle becomes the layer known as the **ectoderm of the embryo**, and the roof of the **ventral endocyst vesicle** becomes the layer known as the **endoderm** (or entoderm) of the embryo. Both layers cling together by means of metabolism. Adjacent to the intensively growing ectoderm, the endoderm becomes so stretched that it forms only a single, thinner layer of flatter cells. On the other hand, the taller ectodermal cells stand pressed closely together, phalanx-like. Owing to their elongation toward the free surface at the floor of the amniotic sac, one concludes that these cells are pushing against each other laterally. Therefore, in the biodynamic metabolic fields of the endocyst disc, the cells of the ectoderm constitute a "pushing layer," whereas the endoderm represents a "pulled layer" (Fig. 2.8ff.). The following statement is a valid rule: *Wherever cells lie close together, extended perpendicular to a free surface, they exert a mutual pressure in a lateral direction.*

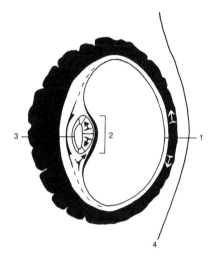

Fig. 2.8. Fully implanted conceptus, about 7 days old, approximately 0.3 mm in greatest diameter. Toward the end of first week, the conceptus has two chambers. The ectoblast (or trophoblast) is indicated in black and everything inside the ectoblast is defined as endoblast. Converging double half arrows: restraining function of inner tissue arising at interface between trophoblast (ectoblast, black) and endoblast. Diverging black double arrows with cross-tails: surface growth of ectoderm. Ectoderm (hatched) is covered locally by endoderm (black). Tailed arrow: incipient nutrient-transport through anlage of body (connecting) stalk. The blunt projections at outer margin of ectoblast represent solid columns of cells of ectoblast (primary chorionic villi). Diverging white double arrows with cross-tails: surface growth of ectoblast. 1) anlage of yolk sac, 2) endocyst disc (ectoderm and endoderm), 3) roof of developing amniotic sac.

Third Week: Development of the Three-Chambered Conceptus

In contrast to the vigorous volume growth of the ectoblast, the inner cells of the conceptus (**endoblast**) initially grow quite feebly. As a consequence of rapid differential growth, the ectoblast glides away from the endoblast and an intermediate layer of loose tissue forms between them. It is as if this loose tissue constituted a compensating layer arising through a process resembling **dehiscence**. It is probable that many of the cells in this intermediate layer are simply left behind by the faster growing ectoblast as it lifts away from the endoblast. From a biomechanical point of view, the tissue in this layer is strained under tension in circular and radial directions as the conceptus enlarges (Figs. 2.9, 2.10). As far as the cells of this intermediate layer are concerned, they become flatter and this leads to a loss of their intracellular fluid. This fluid collects together in the **interstices** as intercellular substance. In this way the tissue becomes reticulated or honeycombed. The network is the middle blastocyst layer, the so-called **mesoblast** (Fig. 2.9).

With further growth of the conceptus, the network of mesoblast cells tears apart, leading to the formation of a new chamber or so-called **chorionic sac** (also known as the extra-embryonic coelom; Figs. 2.10, 2.12). At this stage the conceptus has three chambers. A small amount of mesoblast remains, constituting (i) the **lining mesoblast** at the periphery of the chorionic sac and (ii) the **covering mesoblast** of the endoblast (endoblast + covering mesoblast = endocyst). The transition region between the lining mesoblast and the covering mesoblast is called the **connecting stalk** (Fig. 2.11). The interstices within the remaining mesoblast communicate with each other and fluids start to trickle through them over the surfaces of the conceptus, well before a network of blood vessels arises. The working hypothesis here is that substances from both the yolk sac, as well as from the cells of the lining mesoblast (chorionic cells), are available as food for the endocyst disc. The nutrients from the **chorion** can flow along the connecting stalk of the endocyst into the endocyst disc between the ectoderm and

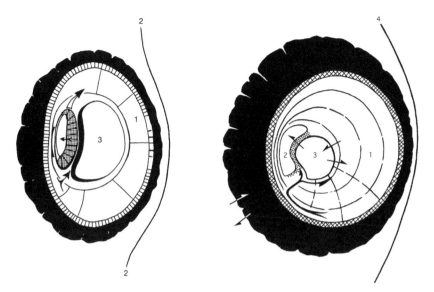

Fig. 2.9. Human conceptus in middle of 2nd week of development, approximately 0.6 mm in greatest diameter (note change in scale from Fig. 2.8). Loosening of yolk sac and formation of mesoblast (1). The radial lines in region 1) indicate alignment of strands of mesoblast cells. The ectoblast is now composed of an outer syncytiotrophoblast (black) and an inner cytotrophoblast (hatched) layer. Outer black: ectoblast. Ectoderm (stippled) plus endoderm (solid inner black) = endocyst disc. Outlined arrow: incipient growth movement of ectoderm at superior end of endocyst disc. Converging double half arrows: restraining function of amnion. Large tailed arrow: transport of nutrients at site of future body stalk. Small tailed arrow: formation of amniotic fluid. 2) luminal surface of uterine mucosa, 3) blastocoele (anlage of lumen of yolk sac).

Fig. 2.10. Human conceptus approximately 0.8 mm in greatest diameter. Progressive detachment of chorion from yolk sac with formation of chorionic sac. Small half arrows: metabolic movements perpendicular to various boundary surfaces. 1) developing chorionic sac (arcs and radial lines represent strands of mesoblast cells; pools of fluid accumulate in their interstices), 2) amniotic sac containing amniotic fluid, 3) yolk sac, 4) luminal surface of uterine mucosa. Large tailed arrow: metabolic movements from the chorion through the body stalk to the endocyst disc (stippled) occurring mainly parallel to the surfaces of the mesoblast. Black: syncytiotrophoblast. Crossed hatching: cytotrophoblast.

endoderm. Thus the pathway of future blood vessels in the connecting stalk is already traced out (Figs. 2.10, 2.11).

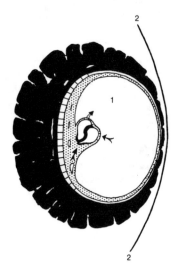

Fig. 2.11. Three-chambered human conceptus at beginning of 3rd week of development; approximately 2 mm in greatest diameter. Black: ectoblast (syncytiotrophoblast). Hatching: transition zone in the ectoblast (cyto-trophoblast). Stipple: mesoblast. The lining mesoblast lines the chorionic sac, the covering mesoblast covers the two-chambered endoblast. The solid tailed arrows represent nutrient-uptake from chorionic sac and body stalk; outlined arrow signifies growth movement of amnion. 1) chorionic sac, 2) luminal surface of uterine mucosa.

Formation of the Axial Process[9]

The endocyst disc is the **anlage of the embryo**, which can be visualized if one cuts away the membranes that enclose the disc (Fig. 2.12). The anlage of the embryo is broad and blunt at one end and narrow and pointed at the other. At no stage is the disc flat, but always exhibits high and low reliefs. The blunt end already indicates the anlage of the superior end of the body, or head region, and conversely, the pointed end is the inferior part of the trunk. The ectodermal surface facing the

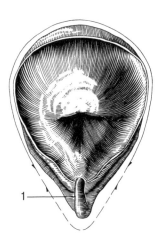

Fig. 2.12. Inner part of the conceptus represented in Fig. 2.11. Human endo-cyst disc 0.23 mm long viewed from dorsal aspect (i.e., looking onto surface of ectoderm) with amnion cut away. Near the center are the expansion dome (light), the rolling rim, and the impansion pit. Converging arrows: growth resistance associated with the tensing and restraining function of the mesoblast along margin of endocyst disc. 1) allantois. (Blechschmidt Embryo, Carnegie Stage 6).

amniotic fluid is the dorsal (back) side of the embryo, and the endodermal side roofing the yolk sac fluid is the ventral (abdominal) side of the embryo. More than half of the endocyst disc represents the anlage of the future head, and above all, of the brain. The predominance of the brain, which is typical for humans, is already apparent at the fourteenth day of development when the endocyst disc is only 0.23 mm long. The neck and trunk region appears to be merely an appendage to the young head region.

The broadness of the head region of the body, when compared to the lower region near the **body stalk**, is a sign that growth at the "free" end of the endocyst disc occurs more rapidly against a lower growth resistance than in the region of the body stalk, where growth is restricted. In the head region, the ectoderm bulges into the amniotic sac and forms the high relief of the so-called **expansion dome**. In opposition to the expansion dome, a depression appears in the trunk region of the embryo, the so-called **impansion pit**.[10] There is a well-defined rim where the expansion dome rolls over into the impansion pit. With surface growth of the expansion dome, this rim rolls more and more over the impansion pit, so that a finger-like **invagination** arises, the so-called **axial process** (Figs. 2.12–2.16). This formation of the axial process has no relation to **gastrulation** occurring in amphibians—human embryos do

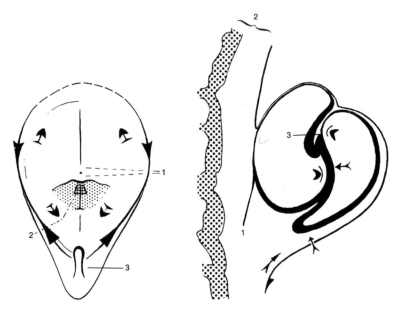

Fig. 2.13. Interpretation of Fig. 2.12. The arrows represent growth forces. Arrows with cross-tails: growth pressure during the surface growth of ectoderm. Converging double arrows: growth resistance of mesoblast along margin of endocyst disc. 1) zone where the ectoderm of the impansion pit has been covered over by the ectoderm of the expansion dome, resulting in formation of axial process (arrow), 2) rolling rim, 3) body stalk (containing allantois). Stipple: impansion pit.

Fig. 2.14. Longitudinal (sagittal) section through the two-chambered endocyst (more detailed depiction of part of Fig. 2.11) showing formation of axial process. Tailed arrows: metabolic movements from yolk sac, from chorionic sac, and through vessels of body stalk, respectively. Arrowheads: fluid pressure forcing the ectoderm and endoderm together. 1) site of future body stalk, 2) chorion with ectoblast (stippled) and lining mesoblast (white) merging with covering mesoblast, 3) apex of axial process.

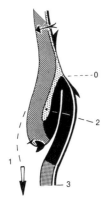

Fig. 2.15. Longitudinal (median) section through subsequent stage of endocyst disc in region of axial process (i.e., in the central zone of the disc). Ectoderm of axial process black, remaining ectoderm stippled. Thin black line with converging double arrows: growth resistance of endoderm (3). Tailed arrow: nutrient-uptake. Arrow with cross-tail: inward movement of ectoderm at the rolling rim. 1) with outlined arrow: growth movement of rolling rim in relation to the nullpoint 0 at apex of axial process. 2) gliding layer between ectoderm and axial process; the mesoderm of the embryo arises in this gliding layer lateral to the axial process.

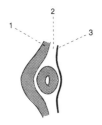

Fig. 2.16. Axial process in cross-section (vicinity of line 2 in Fig. 2.15). The tube-like axial process (stippled) is covered by ectoderm of expansion dome (also stippled). 1) ectoderm, 2) mesoderm, 3) endoderm.

not "gastrulate"! An inconstant aperture at the impansion pit between the dorsal and ventral endocyst vesicles, which has been termed the "neurenteric canal," is sometimes seen in young human endocyst discs; this "canal" is probably an artifact caused by poor fixation or shrinkage during histological processing.

While the rolling rim is always pushing farther over the impansion pit, there is almost no growth occurring within the axial process itself. Therefore, the axial process, which is a structure with a low rate of cell division and very few growth enzymes, elongates but not of its own accord. Rather, the axial process elongates because the ectoderm of the expansion dome keeps rolling over and contributing new cells to its base. Therefore, the apex of the axial process can be considered as the center or, better, the nullpoint for the developmental movements of the whole endocyst disc. The apex of the axial process provides a natural reference for interpreting all subsequent biomechanical movements and the action of biodynamic forces. The lumen of the axial process soon vanishes and what remains is the so-called **notochord**, a thin cord of pale staining cells, that later lies ventral to the spinal cord within the vertebral column.

In the region of the expansion dome, the surface growth of the ectoderm is more intensive than that of the endoderm. The opposing growth of ectoderm and endoderm leads to their gliding apart, and the formation of a loose transition layer between them, the so-called **mesoderm**. The mesoderm cells are left behind by the more rapidly growing ectoderm: this is a process of cell deposition and does not represent an active migration of ectodermal cells into the transition region. The spongy mesoderm is rich in fluid. It is probable that the fluid arises not only from the mesodermal cells that are becoming spindle-shaped, but also from the lumen of the yolk sac and from the margin of the endocyst disc (and therefore from the covering mesoblast), as well as from the chorionic sac (Figs. 2.17, 2.18). The interstices of the mesoderm communicate with each other and so permit an extensive transport of metabolites all along the basal surface of the ectoderm.

The formation of the mesoderm of the endocyst disc between the two gliding layers of ectoderm and endoderm is a repetition of the earlier event seen in the whole conceptus, where mesoblast formed as ectoblast glided away from the more slowly growing endoblast. This is a recapitulation, but it is ontogenetic, not phylogenetic! The formation of the mesoderm marks the transition from an endocyst disc to an embryonic disc.

Fourth week: Formation of the Embryo

Head, neck, and trunk regions: At the beginning of the 4th week of development, in an embryo 1.8 mm long, the head, neck, and trunk regions can be clearly distinguished. The broad head region narrows to a waist-like neck region that passes over into the trunk. In the 1.8 mm long embryo illustrated in Figure 2.17, both amnion and yolk sac have been cut away, leaving mesoblast at the perimeter of the embryo between the amnion and yolk sac. The relatively poor growth of the mesoblastic tissue offers a resistance to the intensively growing embryo

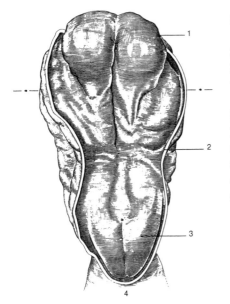

Fig. 2.17. Human embryo about 1.8 mm long, toward the end of the 3rd week of development (Ludwig Embryo, Carnegie Stage 9); viewed from dorsal side, amnion cut away. 1) head region showing neural groove between two dorsal brain bulges, 2) neck region, 3) trunk region. Dot near end of midline groove: entrance to the still-hollow axial process. 4) body stalk. Dot–dash line: plane of section of Fig. 2.18.

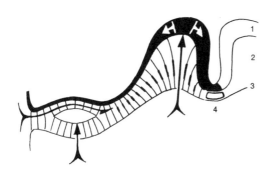

Fig. 2.18. Cross-section of the 1.8 mm long embryo at level of dot–dash line in Fig. 2.17. Ectoderm black, mesoderm hatched (hatching represents alignment of mesodermal cell membranes), endoderm thin black line. 1) ectoderm, 2) mesoderm, 3) endoderm, 4) axial process. Arrows with cross-tails: surface growth of ectoderm at neural crest. Tailed arrows: metabolite transport (metabolic movements) aligned with mesodermal cell membranes. Space in mesoderm at left is part of body sac (intra-embryonic coelom).

and therefore exercises a restraining function. In this way, the embry-onic anlage becomes smaller in the neck region. Meanwhile, at the head end of the body, the mesoblastic tissue appears to cut a deep groove between ectoderm and endoderm, and at the lower end of the body, the mesoblast merges into the connecting stalk (Figs. 2.19, 2.20).

The growth of the embryo is most pronounced in the head region. Dorsal bulges develop on both sides of a zone of near symmetry known

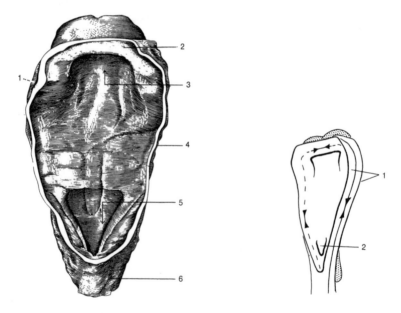

Fig. 2.19. Ventral view (abdominal side) of the same 1.8 mm long human embryo as seen dorsally in Fig. 2.17 (Ludwig Embryo). Yolk sac cut away (white band). 1) location of right-hand entrance into body sac (right coe-lomic portal), 2) thoracic region of embryo near upper rim of umbilicus, 3) entrance to anlage of foregut (superior intestinal portal), 4) neck region of embryo, 5) entrance to anlage of hindgut (inferior intestinal portal), 6) body stalk.

Fig. 2.20. Oblique left ventral view of same 1.8 mm long embryo. Converg-ing double arrows represent restraining function of mesoblast at edge of umbilicus. Ectoderm of dorsal brain bulges and at lower trunk end of body is stippled. 1) mesoblast at umbilical rim, 2) inferior intestinal portal.

as the **neural groove**. The two bulges probably arise from the single expansion dome because the surface growth of the latter is restricted, both at its perimeter and also along the central axis overlying the notochord. As a result of its growth against resistance, the ectoderm buckles longitudinally. At this stage the embryo is still relatively disc-like in spite of its folding. The different forming capacities of ectoderm and endoderm are clearly evident in the correspondingly different relief formations observed on the dorsal and ventral sides of the embryo. The elevated relief in the ectoderm of the head and trunk regions is coupled directly to the formation of a depressed relief on the endodermal (ventral) side. The growing ectoderm draws in the endoderm from around the edge of the **umbilicus**, forming the so-called **superior** and **inferior intestinal portal** (Fig. 2.19). Similar to the form of the entire embryonic anlage, the perimeter of the umbilicus is also broad at its superior end and pointed at its inferior end (Fig. 2.20). The embryo hardly protrudes beyond the edge of the umbilicus. The umbilicus, or ventral aspect of the embryo, remains wide open.

The mesoderm of the embryo arises in a loosening intermediate layer between ectoderm and endoderm. Through its interstices, the mesoderm transports nutrients parallel with, and perpendicular to, the surface ectoderm (Fig. 2.18). As before, the ectoderm at this stage of development is still the active forming apparatus or motor for the folding of the young embryo. From comparisons with later stages of development and with other regions of the body, it is known that no orderly transport of nutrients is possible without the mesoderm.

Body sac, anlagen of the heart, and blood vessels: At the beginning of the 4th week of development, the fluid content of the mesoderm at the superior rim of the umbilicus rapidly increases. This increase in fluid volume leads to the formation of the body sac or **intra-embryonic coelom** (Fig. 2.21). On each side of the embryo, a small opening arises between the body sac and the chorionic sac: the so-called right and left **coelomic portal** (Figs. 2.21, 2.22). The *heart* forms in the dorsal wall of the body sac (Figs. 2.26, 2.27). Initially the heart is represented by a zone of mesoderm between the superior umbilical rim and the intensively growing anlage of the brain. Fluids are already flowing in

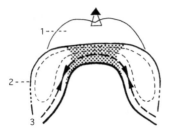

Fig. 2.21. Superior umbilical rim of the 1.8 mm long embryo (part of Fig. 2.19). Heart anlage stippled. Dashed contour: anlage of intra-embryonic coelom. Thick black line: endoderm. Outlined arrow: direction of growth. Converging double arrows: restraining function of the vascular anlagen passing to the heart. 1) dorsal brain bulge, 2 amnion, 3 lateral rim of umbilicus. The dotted ectoderm between 2) and 3) indicates the anlage of the right coelomic portal (communication of chorionic sac fluid with intra-embryonic coelom).

this zone. One can say that the development of the heart takes place to meet the **vascular** requirements of the young brain. The heart is already beating at the start of the 4th week. Even in the 1st month, as the heart is growing to meet the vascular requirements of the brain, so the *liver* is forming to assist the heart, acting as a kind of pre-filter for fluids moving to the heart. The liver is constructed partly from endodermal cells and thus from the anlage of the intestines. These endodermal cells are growing into the loose tissue at the superior rim of the umbilicus (at the **transverse septum**; see Chapter 7).

The first *veins* appear as plexiform spaces in a mesodermal canalization zone along the lateral margins of the embryonic anlage, close to where the ectoderm changes into the amnion. The veins, which convey blood from the vessels of the connecting stalk toward the heart (Figs. 2.22, 2.24, 2.25, 2.26), are the future umbilical veins. From here the oxygen-rich blood is conveyed primarily to the anlage of the brain. From the dorsal brain bulges, blood flows at each side of the neural tube or groove back to the connecting stalk. This pathway is prepared by developmental dynamics as follows: when the neural groove closes to form the neural tube, the mesoderm flanking each side of the tube gains additional room. These lateral zones do not remain empty but

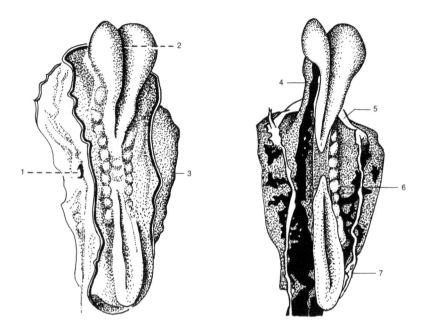

Fig. 2.22. Human embryo 2.2 mm long with 7 pairs of somites, 23 days old (Payne Embryo, Carnegie Stage 10). Left: amnion and yolk sac cut away, dorsal view. Right: portion of ectoderm removed to reveal arteries (black), veins (white). 1) left-hand entrance to body sac (coelomic portal), 2) neural groove already closed to form neural tube in neck region, 3) endoderm, 4) endoderm (concealing heart from this aspect), 5) right venous inflow to heart, 6) somite, 7) right umbilical vein in region of body stalk.

become filled with flowing intercellular fluid, leading to the formation of a pair of vessels, the anlagen of the two embryonic *aortae* (Fig. 2.22ff.). The above development occurs because, even before the actual blood circulation exists, there is already a movement of nutrient-rich intercellular material down a concentration gradient toward the brain. These nutrients move through the interstices of the mesoderm. Vascular formations are therefore a sign of the biodynamic nature of differentiations. It is important to know that the first *blood vessels* are very fine, canal-like routes for the movement of intercellular substance in inner tissue. These canalizations are invariably found where there exist both the spatial opportunity and the appropriate spatiotemporal (physical) conditions for their formation, that is, where metabolic gradients are at work.

Neural tube and somites: The closure of the **neural groove** to form the neural tube is yet another expression of the forming power of the ectoderm. While the thick ectoderm of the neural groove is producing many daughter cells by cell division[11] at the amniotic surface, the offspring are being pushed deeper toward the source of their nutrition, growing as they displace. Here they grow into new cells capable of further division. With the growth of the membranes of these cells, the basal surface of the neural groove enlarges faster than the area of the surface in contact with the amniotic fluid. This differential growth on the two surfaces of the neural groove forces the groove to close gradually to form the **neural tube** (Fig. 2.23). The cause of neural groove closure is therefore a consequence of the global growth of the whole embryo and is not to be sought in the chemistry or structure of individual cells of the groove. The closure of the neural groove begins piecemeal in

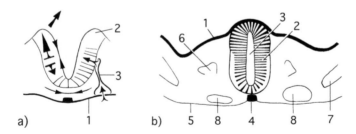

Fig. 2.23. a) Cross-section through 2.2 mm long embryo in region of inferior neuropore. Arrows indicate growth movements associated with closure of the neural groove to form the neural tube. Tailed arrows: nutrient transport. Diverging double arrows with cross-tails: surface growth of ectoderm. Converging double arrows: restraining function in vicinity of free (amniotic) surface of ectoderm and in meninx above the axial process (black). Simple arrow: growth movement during closure of neural groove. 1) endoderm, 2) crest of neural groove, 3) dorsal branch of dorsal aorta. b) Cross-section through 2.9 mm embryo in region of neural tube. 1) ectoderm, 2) neural tube, 3) central canal of neural tube (neurocoele), 4) notochord (axial process), 5) endoderm, 6) somitocoele, 7) intra-embryonic coelom (early pleural sac), 8) dorsal aorta.

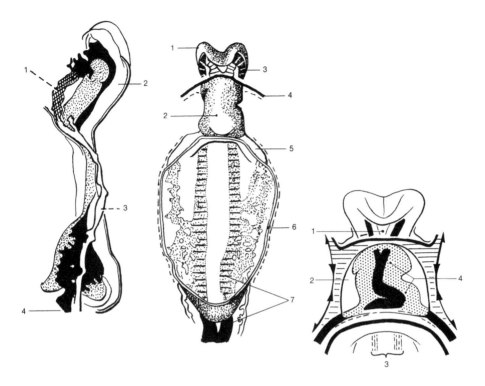

Fig. 2.24. Left lateral view of 2.2 mm long human embryo shown in Fig. 2.22 (Payne Embryo). Endoderm stippled. Heart (1) cross-hatched. 2) neural tube. Umbilical vein (3) white. Ventral aorta conveying blood away from heart, 1st visceral (pharyngeal) arch artery, dorsal aorta, and umbilical artery (4) are all black.

Fig. 2.25. Ventral view of same 2.2 mm long embryo (Payne Embryo). The endoderm is depicted as transparent. 1) neural bulge, 2) heart, 3) 1st visceral (pharyngeal) arch artery (left), 4) cut edge of amnion, 5) left venous inflow to heart, 6) cut edge of yolk sac wall, 7) left dorsal aorta and umbilical artery. Dotted lines: mesoblast.

Fig. 2.26. Thoracic region (superior umbilical margin) of human embryo about 2 mm long, ventral view; compare with Fig. 2.25. Mesoderm hatched; heart wall (cardiac jelly) stippled; blood column in heart, and inflow and outflow vessels, black. Converging double arrow: restraining function of mesoderm. 1) mouth (buccopharyngeal) membrane, 2) thoracic part of body sac (intra-embryonic coelom), 3) superior intestinal portal (with dorsal aortae indicated in background), 4) heart containing blood column.

the neck region. Above and below this level, where the young embryo is beginning to bend, the neural groove retains slit-like openings for several days (superior and inferior **neuropore**).

The paired anlagen of the aortae are initially just like capillaries. If one investigates these vessels from the point of view of their significance for blood distribution in the embryo, then one finds that their initial ramifications are directed to the main consumer or sink of nutrition, namely the neural tube. Here one finds the first dorsal branches of the aorta arising at regular spatial intervals (**metamerically**) from each other (Figs. 2.23, 2.29). These vessels subdivide the mesoderm into individual segments called **somites** (Figs. 2.30, 2.31), which are organs

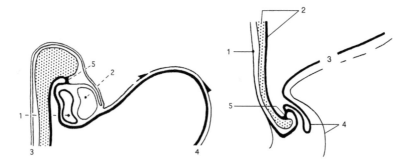

Fig. 2.27. Median view of head region based on serial section reconstruction of human embryo about 2 mm long (schematic). As the yolk sac enlarges, it tilts away from the head; the increasing growth displacement between the yolk sac and the amniotic sac contributes to the formation of the intervening body sac or intra-embryonic coelom. Endoderm, thick black line. 1) heart, 2) body sac (intra-embryonic coelom). The fold where amnion changes to ectoderm lies ventral to the body sac. 3) amniotic epithelium and covering mesoblast, 4) yolk sac epithelium and covering mesoblast. Converging half arrows: relatively slower growth (restraining function) of covering mesoblast. Neural ectoderm stippled. The mouth (buccopharyngeal) membrane (5) is a contact region between ectoderm and endoderm.

Fig. 2.28. As for Fig. 2.27. Lower end of the body. 1) amnion, 2) neural ectoderm (stippled) and endoderm, 3) yolk sac epithelium with covering mesoblast, 4) connecting stalk and allantois, 5) contact region between ectoderm and endoderm (cloacal membrane or anlage of anal membrane).

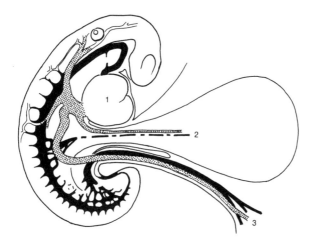

Fig. 2.29. Semischematic drawing based on serial section reconstruction of 2.57 mm long human embryo (Carnegie Stage 12). 1) heart, 2) vessels in yolk sac stalk representing vein (stippled) and artery (black dot–dash line) of yolk sac (vitelline) circulation, 3) vessels in body stalk representing chorionic circulation by umbilical vein (stippled) and arteries (black). Vessels within the embryo represent the embryonic circulation with superior and inferior cardinal veins uniting to form the common cardinal vein that passes to the heart.

of the body wall. The approximate age of the embryo can be determined according to the number of pairs of somites.[12]

The growth of the segmental aortic branches and the subdivision of the mesoderm into somites on either side of the neural tube lead to the formation of indentations or grooves in the overlying ectoderm (Figs. 2.30, 2.33). The ectodermal indentations are a consequence of the fact that so much nourishment is removed from the ends of the aortic branches that their own growth, as well as that of their connective tissue bed (**stroma**), is relatively retarded. Comparative regional investigations have shown that all larger vessels have similar forming functions: they all grow relatively more slowly than the cellular ensembles in their territories of supply.

As stated above, the somites are mesodermal organs. The somitic cells lying near the ectoderm become aligned, initially perpendicularly to the surface of the ectoderm; later they wedge themselves against one another. During their differentiation, a lumen or **somitocoele** arises temporarily in each of the somites. The cells that constitute the floor (medial side) of the somitocoele gradually merge without a sharp boundary into the tissue bed of the neural tube (Fig. 2.32). These cells represent the **sclerotome**, which is the anlage of the axial skeleton (i.e., the vertebral column). With increasing growth of the somitic cells adjacent

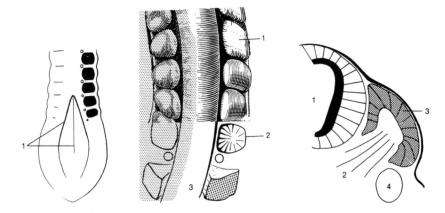

Fig. 2.30. Inferior end of the body of 3.1 mm long human embryo, viewed from dorsal aspect. 1) open neural groove at the lower end of the body (inferior neuropore). Thin strokes signify the shallow indentation (intersomitic sulcus) of the ectoderm, between which adjacent somite swellings (black) arch outwards. Small circles signify vascular branches (dorsal segmental branches of the dorsal aorta) whose terminations cause anchoring of ectoderm of the intersomitic sulci (restraining function of blood vessels).

Fig. 2.31. Part of serial section reconstruction of 2.57 mm long human embryo. Trunk region seen from dorsal aspect. Ectoderm depicted as transparent. 1) dorsal surface of a somite, 2) sectional view of somite indicating alignment of its cells; underlying ring represents dorsal segmental branch of dorsal aorta; below this is a perspective view of positional relation of somite to neural tube (3).

Fig. 2.32. Cross-section of neck region of 3.1 mm embryo represented in Fig. 2.30. 1) lumen of neural tube (neurocoele), 2) sclerotome part of somite, 3) dermatome of a somite, 4) anlage of the embryonic excretory apparatus.

to the ectoderm in the opposite (lateral) side of the somitocoele, the form of the somite changes from an initial vesicular shape to a mushroom shape (Fig. 2.32).

As the neural tube starts to elongate, so the distance between successive dorsal aortic branches also increases. In their turn, the somites therefore become stretched in a direction from the superior part of the embryo to the inferior part (i.e., craniocaudally) and consequently become more and more differentiated. In a stage of development corresponding to Figure 2.34, the somite is seen to have an actively growing layer close to the ectoderm: the **dermatome**. Under this we find a layer whose cells grow slowly and become stretched longitudinally: the

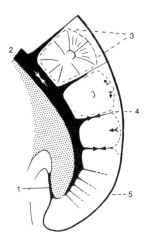

Fig. 2.33. Lateral view of inferior trunk region of 3.1 mm embryo. Endoderm stippled. 1) anlage of anal membrane, 2) aorta (black) with dorsal branches, 3) somite, 4) ectodermal furrow between adjacent somites, 5) mid-dorsal ectoderm over neural tube. Converging double arrows: restraining function of blood vessels. Diverging double arrows: growth expansion of the dermatome. Tailed arrow: release of metabolic by-products into transient lumen of somite (somitocoele) following nutrient-uptake from dorsal segmental branches of dorsal aorta.

Fig. 2.34. Longitudinal section of a somite in neck region of 3.1 mm long embryo. Left: ectoderm, with underlying converging double arrow representing restraining function of basement membrane of ectoderm. Diverging double arrows with cross-tails: growth pressure in dermatome. Converging double arrows above and below somite: restraining function of the segmental (metameric) blood vessels and intersomitic septa. Curved arrows within somite: direction of displacement of dermatomal cells due to growth expansion. Diverging simple arrows at right: growth extension of a myotome.

myotome. The extension of these latter cells conforms to the gradually increasing distance between the segmental blood vessels. The cells of the myotome become the first muscle cells.

In the example of the somite, one sees that apparently essentially different organs (dermatome, myotome, and sclerotome) are really only subtle local modifications of one and the same tissue. This interpretation differs fundamentally from the older anatomical viewpoint, before the concept of regional comparison of organs was discovered, where each organ was considered a specialized entity.

The so-called "gills": A human embryo 2.5 mm long displays characteristic transverse folds between its forehead and its heart swelling (Fig. 2.35). These relief formations are the first facial expressions of the young embryo. In particular, the folds document the growth bending of the embryo: the young embryo flexes itself forward, making, as it were, its first bow. The growth bending has the following dynamics.

The embryonic neural tube, which is the main consumer of nutrients in the young embryo, grows vigorously in length. In contrast, the

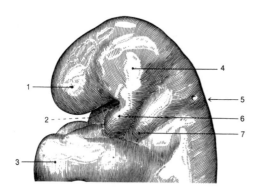

Fig. 2.35. Serial section reconstruction of the head region of a 2.57 mm long human embryo with flexion folds. 1) forebrain region, 2) entrance to the mouth and mouth membrane, 3) heart swelling, 4) upper jaw (maxillary) swelling, 5) hole indicating a narrow passageway to the ear (otic) pit that is transforming to the otic vesicle, 6) lower jaw (mandibular) arch (1st visceral arch), 7) hyoid arch (2nd visceral arch) and inferior to it, laryngeal arch (3rd visceral arch).

growth of the paired aortic anlagen lags behind. These vessels convey much nourishment to the neural tube but retain very little for their own growth. The growth resistance of the aortic anlagen causes the free, flexible end of the neural tube (in the head region) to bend over the heart swelling (Fig. 2.36). This bending leads to the formation of **flexion folds** in the ectoderm. At the same time the embryonic face broadens transversely over the heart swelling. The flexion folds make transverse arches that embrace ventrally the lumen of the foregut as **visceral arches** (pharyngeal arches; see Figs. 2.37, 2.38). The 1st visceral arch is the arch of the lower jaw (mandibular arch), the 2nd is the

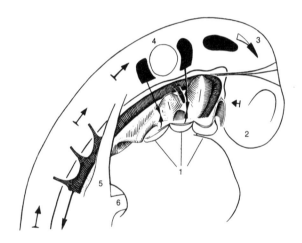

Fig. 2.36. The system of tissue tracts in head region of 2.57 mm long human embryo (23 somite pairs, 26th day). Aorta with 1st and 2nd visceral arch arteries, as well as young nerve trunks (cut away) shown in black. Arrows with cross-tails: growth pressure of neural tube. Outlined arrow: growth movement (flexion) of brain over heart. Simple arrow: inferior continuation of aorta, crossed by superior cardinal vein (5). Converging double arrows: restraining function of connective tissue tracts and visceral arch vessels in flexion folds (visceral arches). Between the folds, the endoderm of the foregut is evaginated as pharyngeal pouches and the ectoderm is invaginated as pharyngeal grooves. 1) 1st–3rd pharyngeal arches, 2) forebrain with eye (optic) vesicle, 3) midbrain, 4) long hindbrain with ear (otic) vesicle, 5) superior cardinal vein, 6) venous junction with heart (sinus venosus).

arch under the tongue (hyoid arch), and the 3rd and the 4th arches are the so-called laryngeal arches.

With increasing flexion of the embryo, the visceral arches in the head region become broader and gradually the tissue inside the arches becomes tensed and oriented in an arc around the foregut. The tissue that is so oriented becomes a conducting structure for large blood vessels, which represent bilateral shunt pathways between the short ventral aorta and the longer dorsal aorta (anlage of the vascular cage of the foregut, Figs. 2.36, 2.38a). Each **visceral arch artery** causes the body wall to arch locally both externally (into the amniotic fluid) and internally (into the fluid of the foregut). As a result, the body wall becomes thick in the region of the arches (around the arteries) but remains thin between the arches. When cut in section, the wall and vessels present a picture as in Figure 2.38b. The thin zones are seen externally as ectodermal grooves (*pharyngeal grooves*) and internally as endodermal pouches (*pharyngeal pouches*). In the region of the pharyngeal pouches the body wall can become so thin that it ruptures: this occurs regularly wherever the ectoderm and endoderm are pressed so closely against each other that there is no room for nourishing inner tissue between the

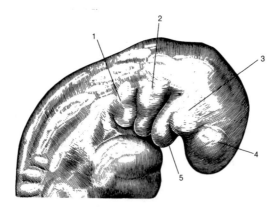

Fig. 2.37. Human embryo 3.4 mm long, 27th day showing flexion folds of head–neck region. 1) 3rd visceral (laryngeal) arch, 2) 2nd visceral (hyoid) arch, 3) upper jaw (maxillary) swelling, 4) eye anlage, 5) mandibular arch (1st visceral arch).

two limiting (boundary) layers. Within hours, the nutrition of the ecto-derm and endoderm becomes so diminished that cells die and defects appear in the body wall. Some children may be born with such defects, described as **congenital anomalies**. Indeed they may remind us of the gills (**branchia**) of fishes or the blowholes of whales, but nonetheless they are not relics from the past as Konrad Lorenz thought. Rather they are occasional phenomena (or borderline cases) that accompany the normal formation of the visceral arches.

Closure of the anterior abdominal wall with the formation of the umbilical cord: The illustrations in Figures 2.39–2.42 show embryos at the commencement of the closure of the umbilicus. If one isolates an embryo about 3 mm long by dissecting away the chorion and amnion

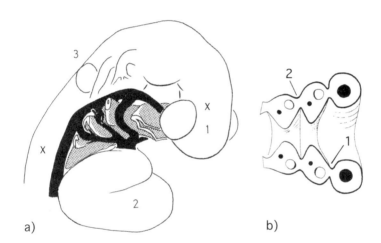

a)   b)

Fig. 2.38. a) Human embryo 3.4 mm long. Vascular cage of the foregut (endoderm indicated by stipple) formed by aortae (black) in head region. The ectoderm is removed. 1) forebrain with eye (optic) vesicle, 2) heart, 3) hindbrain with ear (otic) vesicle. X–X indicates the approximate plane of section for (b). b) Floor of foregut region viewed from dorsal aspect with mouth opening at right. The section is cut approximately in plane X–X in (a). Pharyngeal arches (projecting both externally and internally) con-tain vessels and form external pharyngeal grooves and internal pharyngeal pouches. 1) 1st pharyngeal pouch (right), 2) 2nd pharyngeal groove (left). Aortic arch vessels, black.

(Fig. 2.39), then one observes that the umbilicus on the ventral side of the embryo is still wide open, like a funnel. In the vicinity of the former body stalk, the rim of the umbilical funnel is quite thick and contains the umbilical vessels and the **allantois**. The vessels interconnect the embryo's vessels with those of the chorion. The lumen or coelom of the umbilicus, which is the transition between the chorionic sac and the embryo's body sac, is still wide and contains a connecting stalk to the yolk sac (Fig. 2.40). For orientation, the entrance passage into the right side of the body sac (right coelomic portal) is shaded black in Figure 2.40. The region of the umbilicus remains funnel-shaped, narrowing toward the embryo, until the formation of the actual umbilical cord as follows.

In young embryos, the volume of amniotic fluid is relatively small and the amnion lies close to the surface of the embryo (Figs. 2.27, 2.28, 2.39). With growth, the volume of amniotic fluid increases and the surface area of the amnion enlarges faster than that of the inner wall of the chorion. The amnion approaches the chorion, folding more

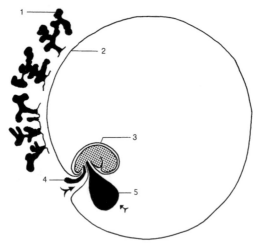

Fig. 2.39. Human conceptus containing embryo 3.4mm long in chorionic sac. 1) chorion with villi, 2) border of chorionic sac (lining mesoblast = chorionic mesoderm), 3) amnion (epithelium and covering mesoblast), 4) allantois (black) in body stalk, 5) yolk sac covered by mesoblast. Arrows represent main directions of nutrient transport.

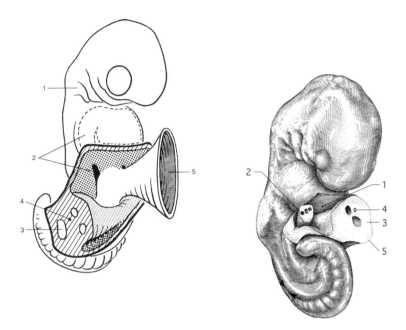

Fig. 2.40. Reconstruction of 3.4 mm long human embryo (about 27 days, Carnegie Stage 12). Umbilicus still wide open (umbilical funnel). Coarse stippling: wall of the umbilical coelom (mesoblast). Thick black line: amnion covering anlage of umbilical cord. Cross-hatching: cut surface of anlage of umbilical cord. 1) 3rd visceral arch (flexion fold), 2) body sac (dashed line) and its right-hand entrance from the chorionic sac (coelomic portal, black), 3) umbilical vein in body stalk, 4) allantois, 5) funnel-shaped transition zone between the yolk sac and yolk sac stalk.

Fig. 2.41. Reconstruction of 4.2 mm long human embryo (about 28 days, Carnegie Stage 14). The embryo is more strongly flexed than 3.4 mm embryo; the umbilicus is more constricted. The umbilical cord is not yet solid, still possessing its lumen (the umbilical coelom, labeled 1). Within this lumen the yolk sac stalk (2) containing the yolk sac duct and the vitelline vessels is cut in cross-section. The locally thickened part of the wall of umbilical cord (3) is derived mainly from the connecting stalk; this portion contains the umbilical vessels (this particular embryo had a normal single umbilical vein, but only one umbilical artery instead of the more usual pair of umbilical arteries). 4) remnant of allantois in wall of umbilical cord, 5) surface of amnion (amniotic ectoderm) covering the umbilical cord. Converging double arrow: restraining function of the dorsal aorta.

and more over the umbilical funnel. The volume of the chorionic sac diminishes, as does that part of it in the umbilical funnel, which still contains the yolk sac stalk (Fig. 2.41). The umbilical vessels remain in the thickest part of the wall of the funnel, which represents the former connecting stalk. In larger embryos (Fig. 2.42), the amniotic sac grows so much that the chorionic sac becomes a relatively small cleft; its lumen begins to disappear and the amnion and chorion eventually fuse together. The umbilical funnel now becomes the umbilical cord covered by the amnion. The cord may contain a remnant of the yolk sac, which tends to be torn away from the intestines already in 7 mm long embryos.

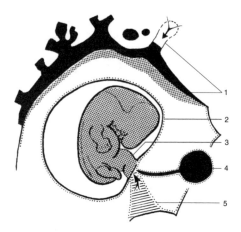

Fig. 2.42. Sketch based on a dissection of conceptus and its membranes, containing 17.5 mm long human embryo, near end of 7th week of development (Carnegie Stage 20). 1) chorionic villus and wall of chorionic sac with ectoblast (black) and lining mesoblast (dotted), 2) amnion with covering mesoblast (dotted), 3) umbilical cord covered by amnion, 4) yolk sac and stalk with covering mesoblast (the lumen of the yolk sac is now disconnected from the gut lumen; the yolk sac remains joined to the embryo via vitelline vessels and so acts as an endocrine gland), 5) body stalk (cross-hatched) of mesoblast. Arrows indicate main routes of nutrient movements.

Chapter 3

Metabolic Fields

A ll anatomical preparations, whether macroscopic or microscopic, are static representations of the body. On the other hand, the organs of the living body are not static but fields of metabolism. These fields are initiated and sustained by living cells. Although almost all cells of the body have the same genetic material in their nuclei, it is apparent that cells grow and develop differently. For example, glandular cells develop differently from muscle cells, and cartilage cells differently again from skin or nerve cells. How does this occur?

In Chapter 1, the principle was proposed that the positional development (topogenesis) of a group of cells is an important prerequisite for their structural differentiation (tectogenesis). For any ensemble of cells, positional development determines development of their form, and this, in turn, determines their structural development. Positional, formal, and structural development go hand-in-hand and it is only in harmony that they bring about development in general. The submicroscopic components of these processes are movements that we call **metabolic movements**. The metabolic processes occurring in cells or their ensembles not only have a chemical significance, but also always display accompanying physical and spatial (morphological) characteristics. Accordingly, one can think of cellular ensembles and organs respectively as locally modified force fields. Here we mean fields that are everywhere impregnated with submicroscopic particles moving in an ordered manner.

From time to time in the past, *biological fields* were invoked to account for ontogenetic differentiations, although such fields were never defined originally as having a kinetic aspect.[13] Today it is known, especially from the studies based on the Blechschmidt Collection of Human Embryos and on the total reconstructions made in Göttingen, that all **metabolic fields** contain metabolic movements that are ordered submicroscopically. The fields can therefore be identified with spatially definable territories of metabolic movements. In the static anatomical or histological preparations of embryos, the fields may be recognized according to the manifold variations in the position, form, and structure of cells and cellular ensembles.

Movements in a metabolic field are a fundamental characteristic of the process of development. The movements of particles in these fields always occur against resistance on the part of their surroundings and thus represent real work in a biophysical sense. This work, when expended over a particular period of time, in turn signifies a particular (biophysical) power, which represents an embryonic **performance** or achievement. This means that the development of a human being, from the earliest stages onwards, can be interpreted, in a dynamic and a biological sense, as a performance specific to the individual.

Limiting Tissues and Inner Tissues: A General Description

In any tissue, all cells are always linked kinetically to one another through the movements of materials. Cells absorb nutrients from interstices and from neighboring cells and, by means of this material uptake, exert a reciprocal attraction to each other. On the other hand, they also exert a mutual repulsion through the release of metabolic by-products. The changing interplay between uptake and release, between attraction and repulsion, is the precondition for cells ordering themselves in particular arrangements and maintaining certain forms.

Even in the earliest stages of development, one finds two characteristically different tissues: **limiting tissue** and **inner tissue**. The former is the boundary between fluid on one side and inner tissue on the other, whereas the latter is enclosed on all sides by limiting tissue and is therefore permanently inside the body. In **histology**, limiting tissue is commonly called **epithelium** and many derivatives of inner tissue are identified as **connective tissue**. Inner tissue could therefore be described as undifferentiated connective tissue. Corresponding to their different locations, limiting and inner tissue have different significance for growth.

In this context, an analogy to the geographical differentiation of various cultures is immediately apparent. Typical regions where civilizations first arose were the banks of fertile rivers, the rich vegetative zones adjacent to river courses and at river deltas, the arable shores by lakes and oceans, and similar habitats. One can think of the ancient civilizations of Babylon, Egypt, Greece, Asia Minor, or Mexico. In comparison with coastal regions, continental inland areas and the central steppes remained culturally undifferentiated for a long time. In these civilizations, it is as if metabolic fields existed where differentiations took place. Whatever products were developed in coastal regions could be advertised through shipping routes, offered for sale, exchanged, transformed, as well as used and improved according to the fluctuations of supply and demand. Frequently these activities were a source of new initiatives and differentiation.

A similar interpretation applies to anatomical differentiation. We can always identify limiting tissues and inner tissues. The limiting tissues (epithelia) constitute the cell mosaics along fluid boundaries. Limiting tissues perform the initial work in the construction of form. They are found at the interface between fluid compartments on the one hand, and inner tissue on the other (e.g., Figs. 3.1, 3.19). A limiting tissue absorbs nutrients from the underlying inner tissue and releases catabolites into the free intercellular fluid. According to the utilization

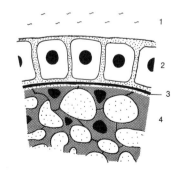

Fig. 3.1. Schematic diagram of the two basic tissues: limiting tissue and inner tissue. 1) fluid, 2) cells of limiting tissue, 3) basement membrane, 4) inner tissue. Close stipple: glycocalyx. Loose stipple: intercellular material of inner tissue.

63

of nutrients, the growing limiting tissue induces oriented metabolic movements from the base of cell toward the free surface and, simultaneously, material movements in the adjacent stroma in the vicinity of the cell base. This leads initially to a flow of nutrients in the interstices of the inner tissue (stroma), and thus to a delicate **canalization**. As discussed in Chapter 2, these tiny intercellular canals are the precursors of subsequent blood vessels, and the concentration gradient of nutrients is an important dynamic factor in the emergence of blood flow.

Limiting tissue can easily unburden itself by releasing its metabolic by-products at one surface into the surrounding fluid. Therefore, there is no opportunity for an interstice, which may happen to exist between some epithelial cells, to become congested with intercellular substance. Thus, a morphological characteristic of all limiting tissue is the formation of a layer of closely packed cells with very narrow intercellular clefts. Limiting tissues regularly consist of many cells and very little intercellular substance. On the other hand, **inner tissue** is surrounded on all sides by limiting tissue and is therefore, as it were, "land-locked." Here catabolites become congested as a distinct intercellular material or **ground substance**; this is a characteristic feature of all inner tissues. As a rule, limiting tissue displays an intensive growth along its free surface in the form of surface (or areal) growth. On the other hand, inner tissue exhibits greater volume growth, so that here one encounters fewer cells per unit volume and proportionately fewer growth enzymes compared to limiting tissue. On account of this, an inner tissue grows more slowly and so generates a (growth) resistance against the surface growth of its overlying limiting tissue.

Previously, the limiting tissue of an organism was usually interpreted as protective, covering structure on account of its topographic relation to the underlying supportive tissue (*epi* = lying upon). In this old interpretation, no notice was taken of the fact that each epithelium usually has two sides. Only one of these is covering; the other has a broad interface with fluids. Interpreted in this way, epithelia are really **diathelia** (*dia* = across) or thin, filtering layers that permit the permeation of materials *across* their surfaces. It used to be said that the **epidermis** (the cell-rich external layer of the skin) was a covering layer and therefore existed essentially for the protection of the underlying inner tissue. However, if this were really the case, then one would expect that the most exposed parts of the body, such as for example the ears and the tips of the nose and fingers, would be protected by an especially thick epidermis. However, this is not the case. Furthermore, in the embryo, many exposed skin territories, such as the skin over the heart-bulge, liver, or brain, are remarkably thin, much thinner than the epidermis on the flanks of the embryo.

Limiting Tissues

It is possible to distinguish thick and thin types of **limiting tissue**. One usually encounters thick limiting tissue in places where the surface growth of the tissue is impeded, and conversely, the thin type in regions where surface growth is facilitated. For example, even in a 20 mm long embryo, the surface growth of the epidermis of the hand-plate is hindered during the formation of the palm. Therefore, the epidermis is much thicker on the palm than over the back of the hand, which has a relatively larger surface area. The callused epithelial thickening, a characteristic of the gripping surface of the adult hand, is already apparent during embryonic development.

A region containing both thick and thin limiting tissue is the epidermis of the embryo's head. Here we find that the epithelium (i.e., the epidermis) over the rapidly growing brain is quite thin whereas, in the vicinity of the flexion folds of the face, it is very thick (Fig. 3.2). During

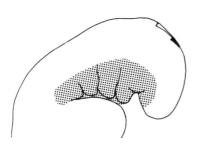

Fig. 3.2. Growth flexion of 2.57 mm long human embryo. The arrow designates the bending of the brain over the heart during the growth elongation of the whole embryo. In the region where surface growth is unhindered, the ectoderm remains thin (upper region). Conversely, wherever surface growth is impeded, the ectoderm thickens (stipple; part of ectodermal ring).

the subsequent stages of development, the packing density of the cells in the underlying inner tissue increases in step with the thickening of the epithelium. With this thickening, the nutritional requirements of the epithelium are increased and the underlying vascular network becomes more closely woven. The striking suffusion of blood to the face of adults is already initiated by the embryo's development.

An analogy from everyday life may illustrate these circumstances. In a city, wherever there is little space available for the construction of houses, wherever construction sites are expensive, the preference is to build multistoried dwellings. The resource consumption of the inhabitants per unit area of construction site is high. The turnover, namely the "metabolism," occurring in the tall buildings necessitates a great supply of consumer goods and an adequate canalization system for the sewage. The converse applies in the country where building space is readily available: here the construction of flatter dwellings is more common.

As a rule, wherever a territory of skin in the adult is normally thick, one finds that the surface growth of that limiting tissue had been restricted during the embryonic period.

An example of "multistoried" epithelial construction resulting from a restriction in surface growth is illustrated by the epidermis shown in Figure 3.3. The most intensive growth of limiting tissue occurs in the vicinity of the nourishing stroma. Here, in the deepest layer of the thick limiting tissue, the cells with restricted surface growth form wart-like pegs that grow into the underlying inner tissue. On the other hand, cells on the free surface receive little nourishment and so this layer grows relatively passively; the poorly nourished layer becomes stretched flat and gradually tears away in small pieces, like the bark of a growing tree. The epithelial cells fall off as keratinized (**cornified**) scales or squames.

In contrast to thick epithelia, thin epithelia grow in such a way that they often remain single-layered (Fig. 3.4). Thin epithelia become stretched over a relatively short time interval and are only slightly hindered in surface growth. Thin epithelia occur quite frequently in the embryo due to the rapid growth of highly cellular, underlying organs such as the brain, the liver, and the heart. At every site, a thin epithelium is a true diathelium, that is, a layer of cells that permits substances to pass perpendicularly across the layer (permeation).

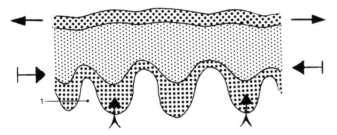

Fig. 3.3. Thick epithelium (in vertical section) with two different boundary layers. The arrows with cross-tails represent a resistance to surface growth, caused mainly by slower growth of the inner tissue. Through uptake of nutrients from the stroma (tailed arrows), the inner boundary layer thickens. The outer more poorly nourished boundary layer grows more slowly (stratum corneum). Light gray stipple: transition layer between deeper well-nourished layer (stratum germinativum) and stratum corneum. 1) connective tissue papilla.

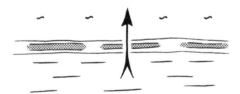

Fig. 3.4. Thin epithelium as a limiting tissue between a fluid (above) and a stroma (below). The tailed arrow indicates that material movements can occur easily through the thin epithelium (diathelial function).

Wedge Epithelia

As a rule, the boundary membranes of cells in a limiting tissue join the inner and outer surfaces of the layer by the shortest possible route. Thus the lateral boundaries of the cells are more-or-less perpendicular to their bases. Now flat surfaces occur almost nowhere in the human body. This means that the epithelial cells almost always have a wedge shape.

Limiting tissues are therefore **wedge epithelia** (or cuneiform epithelia). In reality true cuboidal or columnar epithelial cells, as often schematized in textbooks, are seldom found in the body.

Epithelia forming a free surface that curves convexly outwards consist of cells whose lateral boundaries diverge toward the free surface: these are called *diverging wedge epithelia* (Fig. 3.5). Here the area of the free surface of the cells is larger than the area of the surface facing the inner tissue. In the opposite case of concave-out curvature, the lateral cellular boundaries converge toward the free surface. This arrangement of cell boundaries characterizes the so-called *converging wedge epithelium* (Fig. 3.5). In this case, the area of the free surface is less than that of the surface contacting the inner tissue (stroma).

The shape and disposition of a wedge epithelium have unique significance for the surface growth of the limiting tissue. An example of a wedge epithelium converging to the free surface is the lining epithelium of the young intestine (Fig. 3.6). The growing multiplying cells exert a laterally directed growth pressure on one another. With the aid of this growth pressure, despite the resistance of the adjacent stroma, the surface area increases and thus the intestinal lumen widens.

A characteristic example of a diverging wedge epithelium (diverging with respect to the free surface) is the epithelium at the distal ends of

Fig. 3.5. Wedge epithelia between fluid (wavy lines) and stroma (dashes). Left: (to the free surface) diverging wedge epithelium. Right: (to the free surface) converging wedge epithelium.

Fig. 3.6. Growth dynamics of converging wedge epithelium in a young intestinal tube. Tailed arrow: nutrient-uptake. Diverging arrows with cross-tails: growth expansion of the intestinal epithelium. Outlined arrows: movement of epithelial cells. 1) basement membrane, 2) blood capillaries. Dashes: stroma.

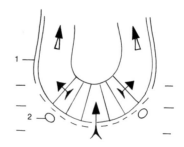

the embryonic anlagen of the limbs (Fig. 3.7). Here the growth of the cells also leads to a growth pressure. This can be recognized morphologically by the local acceleration of growth at the free edge of the limb. As before, the growth pressure results in an increase in surface area. This increase proceeds in the direction of least resistance, as does each growth process under its particular conditions. For the limb anlage, this increase in surface area means that the limb grows and elongates at its apex. A decisive contribution to this process is the growth work performed by the diverging wedge epithelium. Thus all the cells of the young limb do not grow equally rapidly: those at the root of the limb grow slowly and those at the free end grow intensively. The growth at the free end is called **appositional** growth.

A third, special example of wedge epithelium is found in the so-called floor plate of the neural groove (Fig. 3.8). Here the epithelium is of the converging wedge type with the "free" surface bordering the

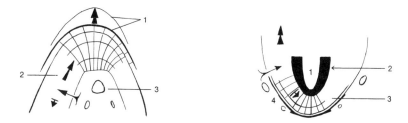

Fig. 3.7. Growth dynamics of diverging wedge epithelium at the apex of a young limb fold (longitudinal section). Tailed arrow: nutrient-uptake. Diverging arrows near (2): surface growth of the ectoderm (2) mainly in the direction of least resistance, which is toward the free end (1). Simple arrow above: direction of limb growth. 3) stroma with blood vessels.

Fig. 3.8. Growth dynamics of converging wedge epithelium in the basal region of the neural groove/tube (cross-section). Tailed arrow: nutrient-uptake. Arrow with cross-tail: growth pressure of neural ectoderm against resistance. The growth resistance is indicated by converging half arrows in the basement membrane. Simple arrow: main direction of growth of neural epithelium. 1) neurocoele fluid, 2) layer of neural epithelium with so-called ventricular mitoses (anlage of ventricular zone), 3) layer of neural epithelium with growing cells, 4) stroma.

amniotic fluid (dorsal endocyst fluid). This epithelium performs work against an especially high resistance from the adjacent stroma and simultaneously against the resistance of the laterally adjacent epithelial cells. The basement membrane is very thick at the floor plate (Fig. 3.8, converging double arrow). Despite their attempt at growth expansion, the converging wedge cells here can hardly overcome the growth resistance. They remain wedged together in a tight corner and retain little intracellular fluid. Their capacity for cell division and, accordingly, their surface growth has already died away by the 2nd month of development.

Inner Tissues

Inner tissue arises in metabolic fields adjacent to all limiting tissue. In inner tissue, the cells move farther and farther away from one another, so that a net-like (reticular) arrangement is produced. Much watery, intercellular substance accumulates in the interstices between the cells, loosening the cellular ensemble. At any stage, loose inner tissue is invariably the result of tissue *becoming* loosened. Examples of this are the mesoblast of the conceptus, the mesoderm of the endocyst disc, and the **mesenchyme** of the embryo. The intercellular substance in fields of loosening tissue contains watery by-products of metabolism (Fig. 3.9). The fluids in the **vacuoles** of the interstices, which are initially solitary droplets, soon coalesce. On account of the initial drop-like shape of the vacuoles, the adjacent cells have concave surfaces (Fig. 3.10). These concavities signify that the hydrostatic pressure in the vacuoles is greater than the net pressure within the cell cytoplasm. In turn, it can be concluded that the cells of the ensemble will become gradually separated from each other in a divergent manner, often only retaining contact at their apices. We do not know exactly by what kind of forces the apices of neighboring cells are held together, although tissue culture provides an opportunity to study the apices in motion. Here they can be observed detaching from the apices of neighboring cells and moving in an amoeboid fashion. No such free mobility can occur in epithelial

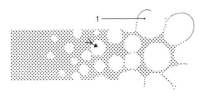

Fig. 3.9. Loosening field. Tailed arrow: exudation of fluid from cells. 1) intercellular substance in interstice. Stipple: collective intracellular material (individual cells not indicated).

LOOSENING FIELD

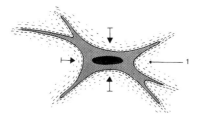

Fig. 3.10. Part of a loosening inner tissue. Arrows indicate the effect of an increase in volume of intercellular material (1) on the form of a cell. Black: cell nucleus.

tissues because there never exists a sufficiently large intercellular space to accommodate the movements. Intercellular fluids invariably play a significant role in the shaping of all cellular ensembles.

In the case where intercellular substances consolidate, so-called **procollagen** is precipitated by the inner tissue and this leads to the gradual development of a tension-resistant system of intercellular fibers of **collagen** arranged in **fascicles**. The consolidated intercellular substance, especially in thin layers of inner tissue such as the fasciae, frequently forms lattices of variable mesh-size with highly deformable interstices (Figs. 3.11, 3.12). The size of the interstices is a characteristic of the degree of tissue loosening. A decrease in the size of the interstices leads to a consolidation of the inner tissue.

We now wish to describe the characteristics of some special metabolic fields that are typical of different regions of the body. In so doing, we will see that the developmental movements, which occur in an ordered sequence in these metabolic fields, always conform to specific biodynamic rules.

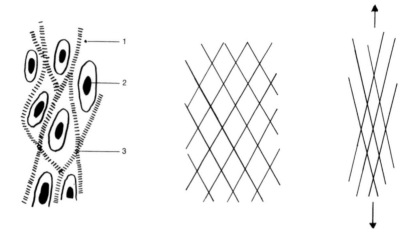

Fig. 3.11. Inner tissue with partially fibrous intercellular substance (cross-hatched). 1) intercellular substance, 2) cell nucleus, 3) collagen fibers.

Fig. 3.12. Different major directions of fibers and the widths of their meshes in tension-resistant intercellular substance.

Corrosion Fields (Fig. 3.13)

The two stick-figures press two walls firmly against each other; the walls are composed of sheets of living cells. At the contact surface, the apposed walls disintegrate and a perforation develops. A similar analogy applies to apposing epithelial sheets. It has already been pointed out that the blood vessels that convey nutrients to, or remove wastes from, limiting tissues run in the underlying inner tissue. Now when two limiting tissues become so closely pressed together that there is no room between them for vascular inner tissue, then the supply of nutrients (or the removal of wastes) is extinguished and the epithelial cells die.

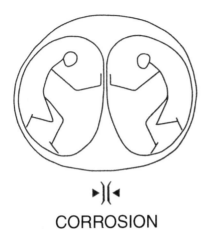

CORROSION   Fig. 3.13. Corrosion field.

Biodynamic metabolic fields, in which epithelial cells die away like this, are termed **corrosion fields**.

One example of the many corrosion fields in the embryo is the developing mouth region. Here ectoderm and endoderm are driven close together, without intervening inner tissue, to form the mouth (buccopharyngeal) membrane (see Fig. 2.27). The single aperture of the mouth arises through numerous small perforations in this particular metabolic field, a corrosion field. Another example of a corrosion field concerns the fate of the two **capillary-like** aortae in the young embryo. These two vessels fuse into one when their thin epithelial walls (called endothelium) become pressed so firmly together by growth and increasing blood pressure that the cells in the contact region disintegrate. In a similar manner, the initially paired ventral spinal arteries become a single ventral spinal artery (Fig. 3.14). Likewise, the kidney tubules, which are initially blind-ended structures containing urine, gain access via corrosion fields to the pelvis of the embryonic kidney. Only then can urine flow into the renal pelvis. If this corrosion process does not take place due to some pathological change, then a so-called urinary cyst will arise. The developing tubules of the cystic kidney become vesicular as urine congests inside them. The so-called cloacal membrane, which is the anlage of the anal membrane (Figs. 2.28, 2.34), is another site of a corrosion field.

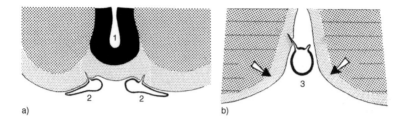

Fig. 3.14. Example of a corrosion field arising between two capillaries that back onto each other in ventral region of spinal cord. a) Spinal cord of younger embryo in cross-section with central canal (1) surrounded by ventricular zone (black), 2) ventral arteries of spinal cord (still paired). b) Older embryo after corrosion field between medial walls of the two vessels has led to formation of an unpaired ventral spinal artery (3) in ventral median fissure of spinal cord. Arrows indicate preceding developmental movements.

Suction Fields (Fig. 3.15)

The stick-figures pull apart a bellows, causing reduced pressure or suction within the bellows. Zones adjacent to limiting tissues, in which suction arises during growth, are known as **suction fields**. As they are located adjacent to a limiting tissue, these zones are also known as **parathelial loosening fields**. Suction fields have various causes. For example, they can arise when the more rapidly growing epithelium tends to lift away from the more slowly growing stroma. This causes fluid to flow from the surroundings into the suction field (Fig. 3.16). In such a region of tissue loosening, wedge-shaped epithelial cells can sprout from their neighbors, absorb materials from the surrounding fluid, and, in so doing, gain extra space for further growth. We recognize such epithelial sprouts as the anlagen of *glands*, both **exocrine** and **endocrine glands**.

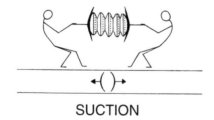

SUCTION

Fig. 3.15. Suction field.

A beautiful example is the formation of glands in the lip. As the lip becomes thicker, so the epithelia of the cutaneous (outer) and the mucosal (inner) side of the lip move away from each other. This leads to the formation of a suction field across the whole thickness of the lip. Sprouts grow from the mucosal epithelium into the zone of suction and become the hose-like tubules of labial salivary glands. The ends of the epithelial sprouts thicken into flask-like (**alveolar**) structures (Fig. 3.16). Here the wedge-shaped cells, which are almost conical, absorb nutrients from the stroma and thereby wedge themselves apart, so that a lumen develops. They then secrete fluid into this lumen. Increasing growth of the wedge epithelium further widens the lumen of the gland. As the gland grows, it pushes away the surrounding inner tissue, which becomes more compact, and consolidates into a glandular capsule (Fig. 3.17). Similar processes lead to the development of sweat glands on the

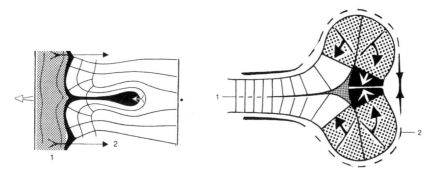

Fig. 3.16. Growth dynamics in a suction field. Formation of a mucosal gland of the lip (gland in longitudinal section). 1) mucosal epithelium, 2) stroma. Outlined arrow: developmental movement of epithelium relative to stroma during growth of the lip. Long tailed arrows: fluid influx into connective tissue, which therefore becomes a zone of loosening tissue. Black: deep layer of epithelium with a growing glandular anlage. Small tailed arrow: uptake of fluid substances through the sprouting region of the gland.

Fig. 3.17. Branching of an epithelial glandular anlage. Coarse stipple: growing wedge epithelium. Black: wedge epithelium hindered in growth. Close stipple: glandular lumen. Arrows with cross-tails: growth pressure of cells. 1) still solid part of gland, 2) arising connective tissue capsule with restraining function indicated by converging double arrow.

external surface of the lip. Wherever glands develop, it is always possible to demonstrate the existence of suction fields in loosening inner tissue. Later in Chapter 7, we will see that the liver and the lung, as well as other organs, are examples of especially large suction fields.

Densation Fields (Fig. 3.18)

The two stick-figures hold a porous dish containing a mixture of solids and liquids. As the liquid drips out, the solid particles sediment and aggregate. In metabolic fields where the biomechanical conditions lead to a loss of water from the intercellular substance, the inner tissue exhibits an analogous thickening. Such fields are called **densation fields**. Densation fields are therefore characterized by the loss of intercellular fluids and the resulting close packing of cells as the tissue consolidates. Densation fields usually arise in deeper regions of the embryo's inner tissue and the small cells, not subjected to pressure or tension in any particular direction, develop rounded shapes as typical **precartilage** cells. All **skeletonization** in the embryo arises in densation fields.

The formation of the skeleton in the arm serves as an example (Fig. 3.19). A densation field arises in an arm during ontogeny in the following way. The limiting tissue (ectoderm) is the major sink of nutrition and procures its nutrients from the underlying inner tissue. In the 2nd month of development, the flow of nutrients to the ectoderm leads to the formation of a dense network of underlying blood capillaries. These nourish not only the ectoderm but also the adjacent stroma. The stroma develops into the highly cellular **dermis** (corium). As a result of the osmotic pressure in its blood vessels, the dermis sucks fluid out

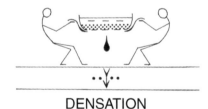

DENSATION

Fig. 3.18. Densation field.

of the deeper layers of the limb. The cells deeper inside the arm anlage therefore become positioned closer to each other as water moves out of this region. This densation field is the zone in which the skeleton of the arm arises. This zone is therefore determined phenogenetically by the skin. The cutaneous anlage of an arm will always lead to the internal formation of a humerus; similarly, under the cutaneous anlage of the forearm, an ulna and a radius will also be laid down. A hand skeleton is never found in the arm despite the fact that the genetic disposition of all its cells, in the form of their DNA, is the same. The subsequent differentiation of a densation field may lead to the formation of not just the cartilage and bone, but also to the ligaments and capsules of the joints, as well as the tendons of muscles.

Wherever the biomechanical conditions in a metabolic field cause the tissue to lose its intercellular fluids so that the cells become consolidated, then a local strengthening of the tissue becomes obvious. In summary, the genesis of a densation field in the interior of inner tissue is a developmental process that is preceded by differentiations taking place further to the outside (*outside–inside differentiation*).

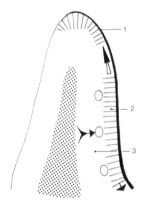

Fig. 3.19. Schematic diagram of early skeleton formation in a densation field. Arm anlage of a 14 mm long human embryo (longitudinal section). Tailed arrow: fluid translocation from the deepest stroma into more superficial blood vessels. Densation field stippled. Diverging arrows: surface growth of ectoderm. 1) epidermis, 2) dermis (corium), 3) hypodermis.

Contusion Fields (Fig. 3.20)

The two stick-figures push together on a springy lattice. The whole lattice narrows and each mesh becomes narrower in the direction of pushing and wider at right angles. Ensembles of rounded cells show comparable morphological changes in contusion fields. Here the cells are pushed together in such a way that they flatten, widening in the direction of least resistance like a compressed rubber ball. The cells become discoidal, which is the typical form of *young cartilage cells* (chondroblasts). In histological sections, the shape of these cells is therefore the sign of this kind of metabolic field. A metabolic field with flattened cells is termed a **contusion field**. Contusion fields develop wherever biomechanical compression is encountered.

An example of such a contusion field is illustrated in Figure 3.21. The diagram shows the side wall of the vertebral canal at the time when the wall is becoming stretched circularly by the growth of the spinal cord together with its adjacent cushion of fluid. The stretched tissue is the so-called "hard skin" of the spinal cord (**spinal dura mater**), which initially is strong only in the ventral region. Further growth of the spinal cord leads to an increase in the tension of the ventral spinal dura. Here, with growth of the spinal cord, the radius of curvature of the dural sheet increases; that is, the ventral part of the dura becomes flatter. This flattening leads to a gradual crowding of the cells on the external (i.e., more ventral) aspect of the dura. The crowding represents a biomechanically compressed zone, that is, a contusion field with discoid cells (young cartilage cells). Such contusion fields also arise at other sites in the body wall, whenever such regions are similarly deformed (see limb development, Chapter 6).

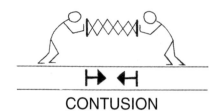

CONTUSION

Fig. 3.20. Contusion field.

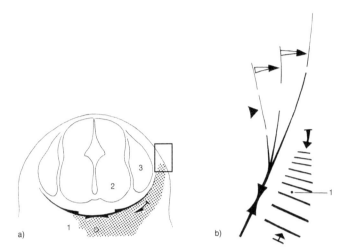

Fig. 3.21. Formation of a contusion field (schematic). a) Cross-section of spinal cord with meninx (anlage of dura) indicated by black line and half arrowheads. 1) contusion field as anlage of a vertebra, 2) spinal cord, 3) spinal ganglion. b) Diagram of boxed region in Fig. 3.21a. Single arrowhead indicates fluid pressure in endomeninx (anlage of arachnoidea). Outlined arrows: successive growth movements of the most recently formed dura as it flattens. Converging double arrows: restraining function of older ventral dura, already stretched and tension-resistant. Converging pair of arrows with cross-tails: contusion field (1) of cells on external aspect of dura.

Distusion Fields (Fig. 3.22)

The stick-figure opposes a resistance by pressing apart two lattices with its arms. In analogy, the growing discoidal cartilage cells push themselves apart (Fig. 3.23). In a biomechanical context, they perform this

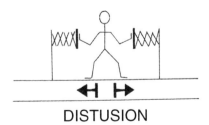

DISTUSION

Fig. 3.22. Distusion field.

function in the following manner: when a contusion field attains a certain magnitude, both the cellular uptake and release of molecular metabolites become increasingly impeded. On account of the intracellular congestion of catabolites, which are of a high molecular weight, the cell develops a high **osmotic pressure** so that water tends to flow into the cell from its surroundings. The young cartilage cells swell, manifesting a so-called *growth swelling*. In so doing, they lose their discoid form and become globular cartilage cells (chondrocytes). By means of their growth swelling, the cartilage cells exert a pushing action in a preferential direction. For the case illustrated in Figure 3.23, the direction of growth pushing is in the long axis of the embryonic finger.

A metabolic field in which growing **cartilage** exerts a pushing function is called a **distusion field**. Such fields have been described classically as zones of chondrocyte hypertrophy. However, the term distusion is used here to contrast the action of growth pushing with contusion,

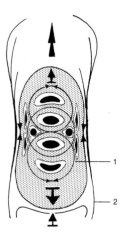

Fig. 3.23. Distusion field in a skeletal anlage (longitudinal section). Growth swelling of cartilage. Cell nuclei black. Gelatinous ground substance capable of swelling, stippled. Fibrous ground substance indicated by lines (1). Restraining function of fibers (i.e., their growth pull) is indicated by converging double arrows. Arrows with cross-tails denote growth pressure. Simple arrow: growth elongation of cartilaginous segment. 2) perichondrium, as well as anlage of joint capsule.

where the cells are growth-pushed. In the German literature, this activity of cartilage cells is known as a *Stemmkörperfunktion*, which can be translated as a "piston-like function." With respect to the piston-like function described in Figure 3.23, the cartilaginous skeleton, and not the musculature, represents the first active component of the organs of movement.

Retension Fields (Fig. 3.24)

The two stick-figures pull on a strong cord so that it undergoes **strain** and becomes taut. The cord does not break and is scarcely deformed. Once taut, the cord offers greater resistance to the pulling figures than, say, a compliant rubber band. In a biodynamic sense, similar characteristics are displayed by inner tissue that becomes stretched and tensed during development (Fig. 3.25). Tensed inner tissue arises in so-called **retension fields**.[14] The following rule is valid: an ensemble of maturing inner tissue cells, whose growth in a particular direction is slower than the growth of adjacent tissues, brings into effect, through its slower growth, a biomechanical resistance to stretching. The growth resistance

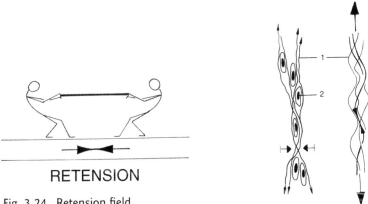

RETENSION

Fig. 3.24. Retension field.

Fig. 3.25. Schematic diagram of a retension field. 1) fibrous intercellular substance in connective tissue, 2) cell nuclei. Arrows with cross-tails: transverse compression. Simple arrows: growth pull. Converging double arrow: restraining function of stretched fibers.

leads to a tensing of the tissue. As a characteristic sign of this tension, the cells in a retension field become spindle-shaped and their nuclei ellipsoidal.

From a large number of examples, one warrants special mention here: the developing heel-cushion (Fig. 3.26). How does the heel develop, apparently so functionally, that it can later serve to cushion the pressure generated during stepping and jumping? As the embryonic skin thickens on the flexor surface of the sole of the foot (corresponding to the stage of development of the hand shown in Fig. 6.26), the dermis becomes richly vascularized. The stems of these subcutaneous vessels become exceedingly strong. On account of its vigorous surface growth, the growing ectoderm of the plantar (sole) skin arches outwards against the resistance of the subcutaneous vessels. Meanwhile the surface growth of the subcutaneous tissue lags behind and this causes the subcutaneous

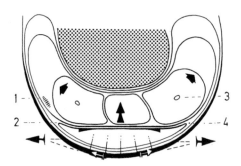

Fig. 3.26. Retension field in heel-cushion of 5-month-old fetus (cross-section). Heelbone (calcaneus) stippled. Arrows with cross-tails: surface growth of epidermis and transverse widening of heel-cushion. Line with converging half arrows denotes connective tissue that becomes stretched, splitting off from the dermis as the heel-cushion widens transversely. Simple vertical arrow: approximation of the deeper stretched connective tissue to the heelbone. Broad arrows: fluid pressure in the lobules of fat tissue. 1) epidermis, 2–4) stretched, deep connective tissue, 3) central blood vessel of a fat lobule. Cross-hatching near 1 denotes subcutaneous (Vater-Pacini) corpuscle.

zone to split into a deep and a superficial layer. The superficial layer remains bound to the ectoderm, and the deep layer becomes stretched and tense. As it splits off and stretches, the deep subcutaneous layer approaches the heelbone and compresses the deeper stroma more firmly against the skeleton. Epithelial sprouts from the ectoderm now grow into the intervening layer loosened by the splitting of the subcutaneous tissue; these sprouts develop into sweat glands (gland formation in a suction field). From a technical point of view, the entire heel (calcaneal) cushion can be compared with the pneumatic principle of the car tire: as the tire is inflated it becomes more firmly pressed to the wheel. As far as the distribution of tension is concerned, the heel-cushion is no more stressed by the voluntary act of an adult's treading than it is in the embryo by the above growth dynamics. The heel-cushion is prepared for its subsequent role by the surface growth of the ectoderm. During embryonic development, the growth of the skin causes the heel-cushion to be pressed against the heelbone, whereas in the subsequent stepping of the adult, the reverse occurs, in that the heelbone thrusts into the heel-cushion.

As a rule, stretched tissue in a retension field functions as a restraining apparatus. All *tendons*, *ligaments*, and *joint capsules* in the human embryo are examples of such stretched tissue. Another example of a retension field is the connective tissue sheath of a blood vessel, the so-called tunica adventitia. The central zone of the embryonic diaphragm is also a retension field; this arises precisely where the growing heart and the enlarging liver move so close together that the intervening connective tissue is compressed while being simultaneously pulled outwards at its periphery. From the viewpoint of developmental kinetics, the fact that ligaments and tendons function subsequently as restraining structures is a result of their being stretched already as they are being formed. The more powerful the retension field in the embryo, the more collagen will polymerize around the stretched cells and the more stretch-resistant will become the tendon or ligament. In this we see a nice example of the frequently occurring reversal between an embryonic growth function and the resulting adult function.

Dilation Fields (Fig. 3.27)

The two stick-figures pull apart an easily extensible structure. In contrast to Figure 3.24, the "band" strained by the pulling yields without offering any significant resistance and so becomes thinner. Analogous biodynamic metabolic fields are known as **dilation fields**. The term dilation is used rather than dilatation, to avoid confusion with the purely mechanical connotation of the latter; in the human body there is no such thing as a purely mechanical process, and therefore no purely physical traction, and also no purely physical extension. However, there definitely exists biomechanical traction and, along with it, the consequences of this traction. Biomechanical traction arises from work performed by living cells and cellular ensembles; we can describe it as an achievement or performance of the whole embryo. Actual biological processes are always more than purely physical events.

In dilation fields, the living inner tissue cells become extended and aligned in bundles (**fascicles**) and sheets. Such cells develop into *muscle fibers*. In a developmentally dynamic sense, the gradual extension of muscle cells is a passive event. Initially muscles are always the more passive component of the *organs of movement*. On the other hand, by means of the growth swelling described above, the young cartilage exerts an active piston-like function, which drags the adjacent cells. If the precursors of muscle cells were not first extended, then they would never be able to shorten subsequently when active. Once again we see a reversal between the embryonic growth function and the adult function. Even adult muscles are still capable of extension: all natural muscular actions begin with a slight extension of the muscle prior to

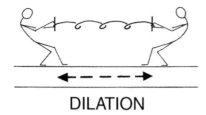

DILATION

Fig. 3.27. Dilation field.

shortening. The interplay between muscular extension and muscular shortening begins very early in human development. Initially, both these actions are growth functions and it is only very much later that they become the signs of arbitrary acts in the context of the voluntary performances of an individual.

In the course of growth extension, the embryonic muscles and muscle cells, as well as their nuclei, become slender. *Muscles* always arise wherever there exist both a biophysical cause (**tensile stress**) and a spatial opportunity; these factors combine in such a way that a small amount of transverse growth always accompanies the longitudinal extension. Wherever transverse growth is impossible and lateral compression occurs instead, then *tendons* arise. For example, everywhere that inner tissue becomes sandwiched between the skin and the broadening articulation head of a segment of developing skeleton, then there is no place for transverse growth.

In dilation fields, the vectors of the longitudinal extension and transverse growth are never constant but change gradually, so that the muscle cells become aligned at an acute angle to their tendons and aponeuroses (Fig. 3.28). The transverse growth of adjacent muscle cells in a dilation field may cause compression of intervening cells, leading to a hierarchy of retension fields that permeate and invest the whole muscle (the various microscopic connective tissues of muscle such as

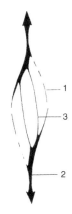

Fig. 3.28. Dilation field of a muscle (longitudinal section). The diverging arrows indicate the main directions of the extending stresses. 1) muscle belly, 2) attachment of tendon, 3) internal connective tissue of muscle (e.g., endomysium, perimysium). Note oblique direction of muscle fibers with respect to direction of extension.

endomysium, perimysium, and epimysium, as well as the fascial sheath of macroscopic anatomy).

Detraction Fields (Fig. 3.29)

The stick-figure on the right pulls a hard support toward itself. The stick-figure on the left tugs on a similar support, which is bound to the first by a medium of viscous glue. The configuration and the glue act in such a way that the supports are pushed closer together as they are dragged. The glue yields, slippage occurs, friction increases, and water is expressed from the zone between the supports. By analogy, meta-bolic fields in which consolidation occurs as the result of water loss accompanying biomechanical gliding movements over a firm pushing substrate are called **detraction fields**. Detraction fields are regions where bone arises (**ossification zones**).[15] This applies to *bone formation* in the matrix of connective tissue and cartilage, as well as to bone formation that continues on a pre-existing bony substrate. Examples are found in the ossification centers of the membranous **calvaria** (intramembranous ossification), in the bony sleeves and layers that arise around growing and dying cartilage cells (perichondral and endochondral ossification, respectively), and in the additional bony growth on previously formed bone (appositional ossification). All these differentiations commence as a consolidation of the tissue accompanied by loss of water and impreg-nation of the intercellular matrix with highly insoluble calcium salts.

Such a metabolic field is shown in Figure 3.30. Here fibrous connec-tive tissue glides over a firm supporting layer, namely the growing

DETRACTION

Fig. 3.29. Detraction field.

terminal segment of the finger (terminal phalanx). As a consequence of the growth pushing (distusion) of the cartilage of the phalanx, the connective tissue at its apex tends to be displaced laterally. Thus the connective tissue here becomes squeezed against the pushing cartilaginous skeleton. With further piston-like growth of cartilage, this connective tissue layer becomes more and more tensed and viscous. Under traction, the viscous connective tissue glides tenaciously over the apex of the phalanx toward the flexor side of the fingers, yielding as it does to the stronger growth pull of the flexors. This gliding leads to the expression of fluid and a hardening of the intercellular ground substance, that is, to a detraction field. Within this field, the end of the young phalanx develops as a bony epiphysis without any prior formation of cartilage, that is, it becomes a site of intramembranous ossification. A considerable time will elapse before the remaining parts of the

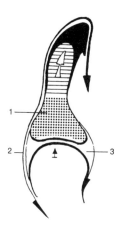

Fig. 3.30. Detraction field at apex of terminal segment (distal phalanx) of cartilaginous finger (longitudinal section). Bone black. Mid-region of terminal phalanx cross-hatched. Outlined arrow: direction of piston-like growth of cartilage. Simple arrow: tissue gliding over growing terminal phalanx in direction of flexor side of finger, resulting in bone formation in a detraction field below. Arrow with cross-tail: growth expansion (of middle phalanx). Half arrows: growth resistance of the cartilaginous "skin" (perichondrium). 1) cartilage forming joint socket (at base of distal phalanx), 2) joint capsule, 3) joint sac.

cartilaginous finger segments become ossified by so-called endochondral ossification.

Another example of bone formation is shown in Figure 3.31. With the elongation of the human face, the distance between the brain and the connective tissue anlage of the cheekbone (zygomatic arch) increases. With increasing distance between the cheek and the brain, the former exerts a growth pull on the stretched membrane around the brain (the dura mater or pachymeninx) that continues superiorly as the connective tissue of the temple region. As a result of this growth pull, the meninx becomes strained biodynamically in a quite specific manner. According to the distribution of tensile stresses (single arrows, Fig. 3.31), the meninx splits into an outer layer that lifts away locally from the inner layer. A narrow, obtuse triangular field emerges: the base of the triangle is the inner layer of dura mater and the two sides

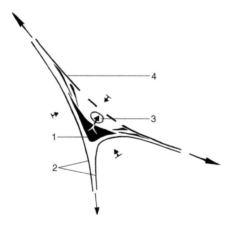

Fig. 3.31. Detraction field in the developing calvaria (right frontal bone) in a human embryo 27 mm long (midline is to right; see Fig. 4.6 for orientation). Simple arrows at top and right represent tensile stresses in meninx covering the brain. Simple arrow at bottom represents tensile stress in cheek region. Arrows with cross-tails: growth pressure of adjacent softer tissues (hypodermis, brain, and zygomatic region). Tailed arrow: release of fluid from extremely compressed tissue. 1) focus of bone-formation (black) in the zone of detraction, 2) connective tissue of anlage of cheekbone, 3) blood vessel, 4) inner layer of dura.

are outer layers of dura. The angle enclosed between the two sides of outer dura was originally close to 180°. As the outer layer yields to the pull in the direction of the lower face, the obtuse angle diminishes and the two adjacent sides of the triangular field come closer. Intercellular water is squeezed from the tissue in the reducing angle, as indicated by the tailed arrow, and blood vessels arise in the vicinity of the base of the triangular field (inner layer of dura) to remove this fluid. The translocation of water signifies a decomposition with local consolidation ("strain-hardening" or "growth-annealing") between the two sheets of the splitting meninx. The consolidated zone represents a detraction field, namely a zone where a center of bone formation arises (ossification center of the frontal bone).

The multitude of possible forms of ossification centers can hardly be reckoned. However, the common principle is that the focus of every ossification zone is a detraction field. These differentiations, too, are living reactions, as are all the processes occurring in metabolic fields.

The symbolic little stick-figures were first used in embryology lectures in the late 1940s when teaching aids were scarce and it was necessary to convey the essence of the new concept of metabolic fields by chalk and blackboard. The stick-figures represent the biodynamic activities of local populations of living cells (i.e., metabolic fields), which in turn affect the intervening or adjacent cellular ensembles in particular ways to make new metabolic fields. On account of their simplicity and comprehensibility, the figures are a most useful aid to learning. We should always envisage the stick-figures, which help us to make the processes of developmental dynamics understandable, as being very much alive; they must be hale and hearty if they are to perform their work!

If the stick-figures were to become ill, if they were to have a genetic defect, or a high fever, or to become infected by a virus, or even to suffer a transient shock, then they might be incapable of a normal execution of the performances described above; possibly they might not be able to perform their tasks at all. This analogy applies to all metabolic fields. It is only under normal predispositions and healthy conditions that the normal developmental movements of cells and cellular ensembles can take place, movements that give rise to differentiations.

Chapter 4

Nervous System

The term "system" implies something contained or delimited. However, it will become apparent in this chapter that there is no sharp boundary between structures and organs that are usually described as "nervous" and other parts of the embryo. Indeed, we develop from one fertilized ovum and it is impossible to say when the first "nervous" cell arises! It is therefore impossible to say when and where the nervous system begins and ends. The same applies to other so-called systems of the body. Nevertheless, long usage has made us more accustomed to the heading of this chapter rather than the phrase "nervous organs."

The Neural Tube

No organ could exist that is functionless during its development. This axiom also applies to the nervous system. The nervous system achieves its subsequent performances on account of its previous growth functions. In investigating the development of the human nervous system, one is astonished to find a unique regularity in the spatial alignment of its cells. In early embryonic stages, this spatial order is so striking that it is evident from the most cursory glance at histological preparations.

As with other organs, the nervous system develops in a dynamic way according to its positional relationships with neighboring structures. As we have seen in Chapter 2, such relationships are already evident for the **neural tube** in an embryo only 2 mm long. Here the wall of the tube is a layer of tall cells extending between the fluid at the inside surface and the vascularized **stroma** at the outside surface. The fluid bordering the inside surface is the anlage of cerebrospinal fluid; so long as the neuropores remain open, it is confluent with amniotic fluid. Only at the outside surface of the neural tube can the cells garner nutrients supplied by blood vessels. It is unknown to what extent these cells might also absorb substances from the fluid within the neural tube. However, in histologically stained preparations, the innermost (ventricular) layer of the neural tube appears almost black in transmitted light because of the high density of cell nuclei present there. Therefore, it is unlikely that major fluid absorption occurs at the inner surface.

Conversely, the outer layer of the neural tube appears pale in histological sections because it is composed of the lightly staining processes[16] of the cells whose bodies (**somas**) are located in the inner layer (e.g., see Figs. 2.23, 4.7). These cytoplasmic extensions lie so close together that initially there is no room in this outer layer for cell nuclei. This particular structural arrangement arises because the somas in the deep layer are striving, by means of their extensions, to obtain nutrients from the stroma over the shortest possible route. These cell processes regularly align themselves perpendicular to the external surface of the neural tube. From a biophysical view, their alignment is a sign of the existence of a laterally directed growth pressure on the part of the tall cells within the wall of the tube. As soon as they absorb nutrients from blood vessels, these cell processes are exerting a lateral growth pressure and are functioning directly to increase the outer perimeter of the neural tube. The processes perform their greatest growth work at the external surface of the tube. If the enlargement of the neural tube were to result from an increasing internal pressure of the fluid in its lumen, then one would expect to find a circular arrangement of cells around the long axis of the tube.

In contrast to the outer layer of the neural tube, the inner layer is largely relieved of performing work. So here one finds a region adjacent to the lumen (**ventricle**) where the cells are continually dividing, the layer of so-called *ventricular mitoses*. As soon as the perimeter of the neural tube begins to enlarge, we find cells being squeezed from the innermost layer to form the first so-called *gray matter* (mantle layer) of the nervous system between the pale and black zones (Fig. 4.7). The gray matter comprises the cell bodies of neurons and **glia**. The outermost layer remains pale as it consists mostly of nerve fibers and other cell processes, the so-called *white matter*. This pattern arises initially in the spinal region of the neural tube, as well as in the head region.

The Central Nervous System: Brain and Spinal Cord

We recall that in Chapter 2 it was shown that the ventral region of the neural tube (where the neural groove first formed) is hindered in its surface growth. Conversely, the neural tube grows faster in lateral and especially dorsal regions, because here growth encounters less resistance than ventrally.

Toward the end of the 1st month, the superior neuropore closes and *brain* growth accelerates with increasingly pronounced flexion of the embryo's back over the heart. The skin of the head becomes stretched and thin over the immense brain. The brain becomes doubled up under the taut skin, so that rather sharply definable segments of the brain become apparent. One observes three large segments that can be likened to the segments of a bent finger flexing against each other. Comparable to the terminal segment of a finger, an initially small *forebrain* arises, followed by a relatively longer *midbrain*, and beyond this, a longer *hindbrain* (Figs. 2.36, 4.13, 4.14). In the 2nd month, the forebrain grows particularly rapidly; its main derivatives are the *cerebral hemispheres* that grow at the sides of the forebrain.

In its development, the brain also follows the general rule that its initial performances are growth performances. These comprise the formation of a highly complex growth architecture with bilateral, almost symmetrical, connections in all axes. The entire, subsequent autonomous activity of the brain is based on these growth connections.

How does it now happen that neural centers, in particular a cortex, are formed? Why do such centers occur only in the brain and not in the spinal cord? What is the basis for the unique differentiation of these centers so that they characterize the human brain? To help answer these questions, it is possible to establish the following facts. At the end of the 2nd month, the cerebral hemispheres have developed an extensive contact surface with the **vascular meninx** (**leptomeninx**; pia mater and arachnoidea). At this time, the surface of the hemisphere is still smooth. Just as occurs for cells in the neural tube, the cells in the wall of the forebrain also obtain their nutrition from blood vessels in the vascular meninx. Thus the processes of these cells become aligned perpendicularly to the surface of the brain. Such positioning, perpendicular to the outer surface, compels us to conclude that the cell processes exert a lateral growth pressure parallel with the external surface of the brain. The basis of this growth pressure is the accumulation of an increasing volume of cytoplasm in the cellular processes.

If one considers that usually there is no biodynamic pressure without an accompanying tension, then it is obvious that we must inquire about structures experiencing tension and about the spatial distribution of both tension and pressure. Actually, one can identify a direct, morphological expression of the distribution of pressure and tension in the brain. All systems in the growing brain that are capable of transmitting or withstanding tension, all cell boundary membranes, are situated with their main alignments either perpendicular to or parallel with the outer surface of the central nervous system. One observes a so-called **trajectorial** construction of the nervous system.

If the embryonic brain is depicted the same size as the adult brain, then the adult brain appears to be shriveled. The surface of the brain remains smooth until approximately the 7th month. Only then will the fetal brain develop its typical furrows and convolutions (**sulci** and **gyri**; Figs. 4.1, 4.2, 4.3). As the cerebral cortex is richly supplied by pial blood vessels, it follows that its surface growth will be very intense. On the other hand the deep layer, or medulla, which consists mainly of the processes of cells and is known as the white matter, has less surface growth. The white matter therefore presents a resistance to the growth of the cortex; the resistance is a tension. This assertion is based on the experience of dissecting fresh specimens, where the fibrous processes are found to be under tension. The membranes of the nerves in the white matter therefore function as restraining structures (Fig. 4.3). Initially, the main alignments of the nerve fibers in the white matter correspond to the expected directions of tensile forces opposing the surface growth

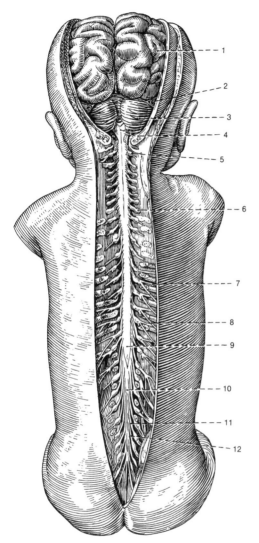

Fig. 4.1. Newborn. Cerebral hemispheres, cerebellum and spinal cord. 1) forebrain gyri, 2) venous sinus, 3) cerebellum, 4) transition region between brain and spinal cord (at obex), 5) and 6) spinal ganglia, 7) and 8) spinal nerves, 9) sacral end of spinal cord, 10) nerve roots in lumbar and sacral regions, 11) and 12) nerve roots with spinal ganglia at lower end of vertebral canal. (after Sobotta and Becker, 1962).

of the cortex. Everywhere in the brain, we find that the neural conduction pathways are aligned in a uniform fashion with their major orientations either perpendicular to or wparallel with the local cortical surface, in conformity with the nature of such traction systems.

Some *nerve fibers* from the brain continue into the spinal cord. As the spinal cord grows in length, so too do these fibers. However, as the fibers grow much more slowly than the cortex, their resistance to traction increases more and more as the cortex enlarges in surface area. In this way, the white matter in the spinal cord plays a role in constructing the relief of the gyri and sulci in the cortex (Fig. 4.2).

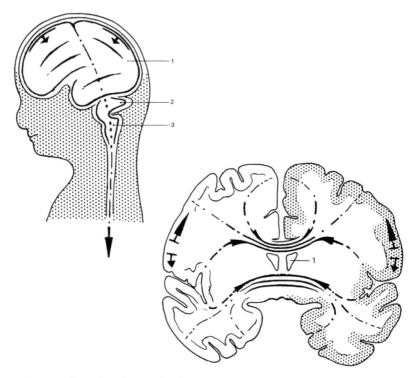

Fig. 4.2. Fetus in 7th month, showing main orientation of forebrain sulci. Arrows with cross-tails: growth pressure of the brain capsule (meninx). Simple arrow: direction of growth tension of the spinal cord. 1)occipital pole of cerebral hemisphere, 2)cerebellum, 3)medulla oblongata with spinal cord below.

Fig. 4.3. Frontal section of human forebrain, representing the growth tension of the white matter, especially the commissures. Right-hand side: cortex stippled, white matter, white. Diverging arrows with cross-tails: surface growth of cortex. Converging arrows: restraining function of the nerve fibers. 1)lateral ventricle of the right cerebral hemisphere.

In contrast to a spinal cord that is becoming hemmed in by surrounding structures, the wall of the brain has the unique spatial opportunity to increase in area. As the surface area of its wall increases, so too does the perimeter of the brain. Dorsolaterally the tissues immediately around the brain and the embryonic skin yield easily to the underlying surface growth of the brain. Bearing in mind that the ventricle is not a site of fluid pressure, the brain also has the opportunity for a growth thickening of its walls along with its surface growth. Investigations have demonstrated that the combination of increase in surface area and increase in thickness leads to the formation of mutually perpendicular interstices in the wall of the brain. The on-going lateral pressure from cells that remain in the innermost ventricular zone permits some cells to slip from the ventricular zone into these interstices. These displaced cells give rise to the *cortical plate*, which is the anlage of most of the *cerebral cortex* or gray matter (Figs. 4.4, 4.5). Similar considerations apply to the so-called *basal ganglia*, which are regions of gray matter deep in

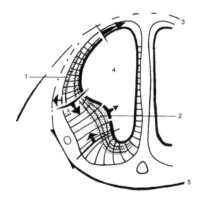

Fig. 4.4. Frontal hemisection of the cerebral hemisphere of a 24 mm long human embryo. Converging arrows with cross-tails: restricted surface growth. Simple arrow with cross-tail: eccentric growth expansion of the brain. Small arrow: direction of growth thickening. Converging half-headed arrows: restraining function of the dura. 1) anlage of cortex (cortical plate), 2) basal ganglia, 3) region of thin skin, 4) lateral ventricle of forebrain, 5) thicker part of the dura and anlage of the basicranium.

the wall of the brain at the base of the cerebral hemispheres. However, here the growth thickening in an outward (pial) direction is prevented by the stretched, tense dura, and so the brain wall increases in thickness by bulging into the lumen of the ventricle (Figs. 4.4, 4.6).

Why does the *spinal cord* lack a cortex? As the vertebral canal remains narrow, the spinal cord has no opportunity for superficial growth thickening through outward cell displacement; consequently, it exhibits no cortical development. It is only the deep regions of the spinal cord that can increase in thickness at the expense of the lumen; *spinal gray matter* is formed here (Figs. 4.8, 4.9). With this increase in thickness, the central lumen or canal of the spinal cord becomes ever narrower. Therefore, the brain grows, so-to-speak, outwards and the spinal cord, inwards.

In the vicinity of the brain, the pia mater and arachnoidea (leptomeninx) enlarge intermittently with each pulse-beat. The yielding pia mater thus provides the brain, to which it is bound, with additional

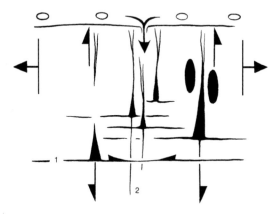

Fig. 4.5. Pyramidal cells of the cerebral cortex (schematic) showing main directions of cell processes at right angles to one another (trajectorial system). Circles: blood vessels of the soft meninx (pia and arachnoidea). Tailed arrow: nutrient-uptake from the pia. Diverging arrows with cross-tails: growth expansion through increase in surface area. Half-headed arrows: major directions of neurite and dendrite growth. Converging half-headed arrows: restraining function of the horizontal nerve fibers. 1) horizontal cell processes, 2) vertical cell processes.

space for surface growth. This surface growth leads to a lower hydro-static pressure within the ventricles of the brain. The weak medial wall of the cerebral hemisphere gives way to this reduced pressure and it

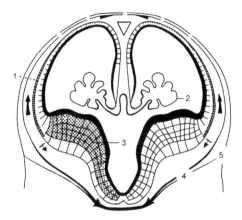

Fig. 4.6. Frontal section through the forebrain of a 27.2 mm long human embryo (somewhat older stage than shown in Fig. 4.4). Gray matter stippled. Diverging tailed arrows: eccentric growth expansion of the brain. Converging arrows: restraining function of the ventral dura. 1) anlage of cortex (cortical plate), 2) invaginated thin medial wall of the brain (choroid plexus), 3) basal ganglia, 4) inner and outer sheets of dura, 5) outer sheet of dura with detraction field in vicinity of splitting of dura (see Fig. 3.31).

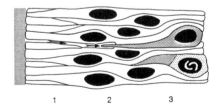

Fig. 4.7. Schematic diagram depicting the mosaic of cells forming in the wall of the brain and spinal cord of young embryos (beginning of 2nd month). Pial covering stippled, at left. 1) white layer (marginal zone, rich in cell processes), 2) gray layer (mantle zone, a zone of cell processes and some cell nuclei), 3) relatively black layer (ventricular zone, rich in cell nuclei and site of nuclear division). Tailed arrows: metabolic movements.

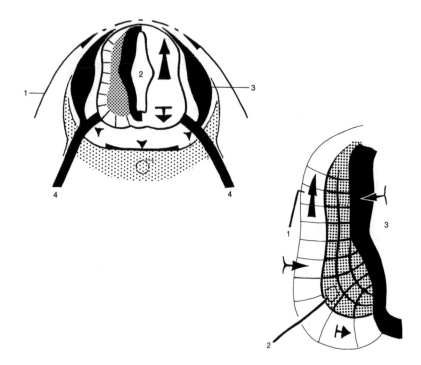

Fig. 4.8. Human embryo 7.5 mm long. Cross-section of spinal cord with spinal nerves. Layer of ventricular mitoses and nerves black, gray matter fine stipple, white matter hatched. Coarse stipple: cartilage. Converging half arrows at top: weak restraining function of the skin over the spinal cord. Two converging half arrows at bottom: strong restraining function of the ventral dura. Diverging arrows on right: growth expansion of the wall of the spinal cord predominantly in a dorsal direction. Arrowheads below: fluid pressure in the bed of the spinal cord (anlage of cerebrospinal fluid). 1) ectoderm, 2) spinal canal, 3) spinal ganglion, 4) spinal (mixed) nerve.

Fig. 4.9. Hemisection of spinal cord in 7.5 mm long embryo. Spinal cord with nerve roots. Layer of ventricular mitoses black; gray matter stippled; white matter light. Diverging arrows: growth expansion through surface growth. Tailed arrow at top-right: possible fluid uptake from the spinal canal by cells during their multiplication; tailed arrow at mid-left: nutrient absorption from pia. 1) afferent pathway (neurite of dorsal rootlet), 2) efferent pathway (neurite of ventral rootlet) arising from cell body in gray matter, 3) fluid in spinal canal. Lines in gray and white matter indicate trajectories of cell processes.

invaginates bilaterally into the ventricle, forming the **choroid plexus** (Fig. 4.6).

Neurons

In anatomy, there is a principle known as **encapsis** or encaptic division, which states that the parts are similar to the whole. When applied to the brain, the principle means that what is valid for the brain as a whole also applies to its individual cells.

A mature nerve cell (**neuron**) has a formal similarity to the entire central nervous system of brain and spinal cord: the cell body (soma) of a neuron corresponds to the brain and the cell process (**neurite** or axon), to the spinal cord. The elongation of neurites represents a significant increase in neuronal surface area (i.e., membrane synthesis) relative to neuronal volume. For synthesis of additional cytoplasm and elongation, neurites require substances that flow mainly away from the cell body (anterograde axonal flow). The passage of these substances into the neurite causes a relative deficiency of material in the soma. Consequently, the soma of the growing neuron tends to collapse and its surface relief changes, resulting in depressions and, as a consequence, elevations in its boundary membrane. The elevations develop into **dendrites** (e.g., see Fig. 4.16). It has been demonstrated that the neurite usually arises first and dendrites later.

How do the complex pathways of neural processes come into existence in the wall of the brain? As a consequence of the combined surface growth and growth thickening of the wall of the central nervous system, the interstices criss-cross each other in principal directions that lie parallel with, and perpendicular to, the pia. Neurites and dendrites grow in these interstices and therefore also criss-cross each other in the same principal directions (Figs. 4.4, 4.5). These directions correspond to the major lines of tension that are to be expected from the increases in area and thickness of the brain wall. It is only at sites where interstitial spaces intersect that unattached cells find sufficient room and have the opportunity to develop into so-called ganglion cells (ganglioblasts; Fig. 4.16 inset). The intersection zones are analogous to the places where street-crossings arise and people may congregate. The cell processes of the emerging ganglion cells conform to this fundamental trajectorial architecture of the central nervous system. In the above way, the "wiring" of the central nervous system arises long before electrical impulses can be detected in the nerve cells. For all these differentiations, it is pointless to invoke a hypothetical "blueprint" in the genes.

The Peripheral Nervous System

This section deals with the nervous elements that anatomy considers as not belonging to the central nervous system. In particular, we answer the following three questions that confront the anatomist interested in development.

How do growth functions determine the number of nerves? In the human embryo, the peripheral nerves arise in definite number and definite sequence. To account for this, it is not necessary to invoke either a mysterious biology of molecules, or habit based on usage in past phylogenetic periods. Rather, we should turn to the biodynamic laws of ontogenetic development (phenogenesis) of the human embryo itself. The first nerves to arise are the so-called **cranial nerves**, then the **spinal nerves**. All the cranial and spinal nerves are present in an embryo when it is about 28 days old and about 4.2 mm long (Fig. 4.11). These nerves develop according to the generally valid rule that for a particular differentiation to take place, there must exist not only a spatial opportunity but also a direct cause for that differentiation.

The investigation of young human embryos shows that the nerves appear regularly for the first time after the dorsal branches of the aortae (dorsal aortic rami) have already reached the lateral wall of the neural tube. These dorsal branches of the aortae function, as do all blood vessels, as restraining structures. They hold on to the growing neural tube just as though they were cords, so that the embryo bends forward as it grows. We can think of the blood vessels as guy-wires that are tethered between the nervous system and the more ventrally located great vessels (Figs. 4.11, 4.12).

Through increase in the number of cells, the spinal cord and the brain are enlarging predominantly in a dorsal direction. On the other hand, parts of the lateral walls of the nervous system remain attached metabolically to the segmental aortic branches. This attachment is already arising before the neural groove closes to form the neural tube. Clusters of nerve cells are thereby drawn out of the dorsolateral margins of the neural groove or tube as the latter glides dorsally away from the blood vessels. The cells arising from the edges of the neural groove and tube in this way are known collectively as **neural crest cells**. Many neural crest cells are the anlagen of **spinal ganglia** (Fig. 4.12). In the head region, neural crest cells can give rise to some ganglia of cranial nerves as well as most of the inner tissue of the head and face (mesecto-derm). The concept that neural crest cells "migrate" of their own accord from the dorsolateral part of the neural groove and tube is incorrect: it ignores the anchoring role of the segmental blood vessels and the dorsal growth and displacement of the remainder of the nervous system. As

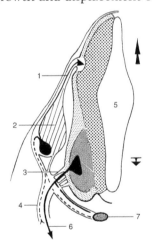

Fig. 4.10. Nerve cells in spinal cord of an 8 mm long embryo. The diverging arrows on the right represent the growth expansion of the spinal cord. Simple arrows at 1 and 6: growth directions of sensory and motor neurites, respectively, laying down the route of the reflex arc. Dashed lines: growth directions of dendrites from spinal ganglion soma. 1) dorsal root fiber, 2) spinal ganglion, 3) ventral rootlet, 4) spinal (mixed) nerve, 5) spinal canal, 6) motor neurite, 7) sympathetic trunk.

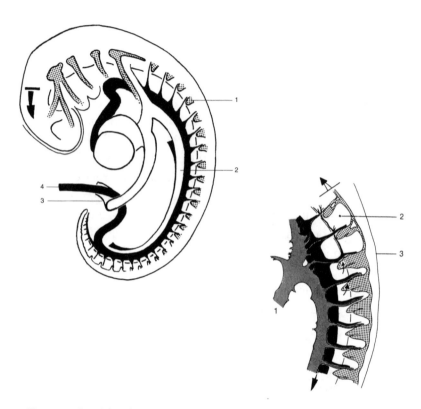

Fig. 4.11. Cranial and spinal nerves of a 4.2 mm long human embryo (Carnegie Stage 13) lateral view. Veins white, arteries black, nerves stippled. Single arrow: growth flexion of the brain. Converging half-headed arrows: restraining function of large ventral blood vessels (aortae, inferior cardinal veins). 1) spinal ganglion, 2) inferior cardinal vein, 3) umbilical vein, 4) umbilical artery.

Fig. 4.12. Part of a serial section reconstruction of a 4.2 mm long human embryo (refer to Fig. 4.11) showing contact zones between neural crest cells and associated dendrites (coarse stipple) and blood vessels. Due to the ascent of the nervous system, each cluster of sensory dendrites lies adjacent to the lower margin of the respective dorsal vascular branch (ramus) of the aorta. Outlined arrow at top: ascent of the nervous system. Simple arrow below: descent of the vascular system. Small arrows: appositional growth of the nerves toward blood vessels. 1) common cardinal vein, 2) location of a sclerotome (focus of segmental skeletonization), 3) dorsal edge of neural tube. Aorta and dorsal rami, black.

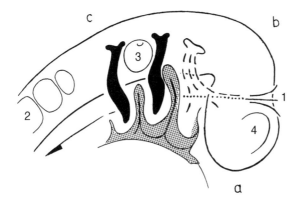

Fig. 4.13. The first sensory nerves in the head region of a 2.57 mm long human embryo, lateral view. The dendrites grow toward the thickened ectoderm of the visceral arches (stippled) that is part of the ectodermal ring. (a) forebrain, (b) midbrain, (c) hindbrain. The trigeminal nerve is indicated by dashes. Nerves of the 2nd and 3rd visceral arches (facial and glossopharyngeal nerves, respectively) black. Half-headed arrow: restraining function of the dorsal aorta. 1) stretched connective tissue extending from dorsal aorta, 2) somite, 3) otic vesicle, 4) optic vesicle.

cells are leaving the dorsal parts of the neural groove and tube from the earliest stages onwards, cross-sections of the spinal cord at later stages will obviously exhibit large *ventral horns* and smaller *dorsal horns* (e.g., Fig. 4.10).

According to the above account of normal development (as experimental investigations of human embryos are impossible), we can now answer the first of our three questions. Normally the same number of spinal ganglia and nerves form as the number of vascular "guy-wires" (connections) to the side of the neural tube. Everywhere that the embryo has greater longitudinal curvature (that is, in the neck region and in the vicinity of the limb anlagen), the nerves that arise at successive distances from each other (metamerically) grow and come into mutual contact. In each of these regions the nerve fibers plait together to form a massive bundle, the so-called **nerve plexus** (cervical plexus, brachial plexus, and lumbosacral plexus; Fig. 4.14). Conversely, in the

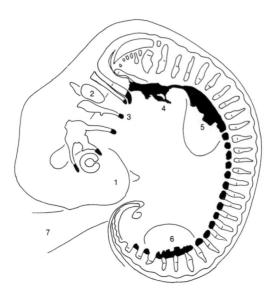

Fig. 4.14. Lateral view of 6.3 mm long human embryo (Carnegie Stage 14) showing plexus formation. Black: the first sites of contact between the (sensory) nerves and the overlying skin, which is here composed of the thickened epithelium of the ectodermal ring (indicated in Fig. 6.2). 1) forebrain, 2) otic vesicle, 3) facial nerve, 4) cervical plexus, 5) brachial plexus, 6) lumbar plexus, 7) umbilical cord.

thoracic region where the embryo is straighter, there is less convergence of the nerves and no significant plexus formation.

What structures become innervated? In young embryos, the most striking innervation by the developing nerves is twofold: (i) thickening epidermis and (ii) muscles arising in dilation fields. Both the thickening epidermis and the developing musculature are regions with special demands for growth. *Thickened epidermis* arises initially in the young embryo only in a zone that extends across the face and downwards over both flanks of the body wall to the lower end of the trunk, then across the groin region. This zone has the form of a folded ring. The skin of this ring-shaped zone is restricted in surface growth and so it thickens, giving rise to the *ectodermal ring* (see Fig. 6.2). Conversely, the skin of the embryo's back (covering the neural tube) and belly (covering the heart–liver mass) is hardly impeded in its surface growth and so becomes thinner in these regions. The first innervation in the embryo arises under the ectodermal ring (Figs. 4.14, 6.2). This innervation, which is the consequence of the formation of dendrites of ganglion cells, already designates those skin regions that later retain an especially rich *sensory* nerve supply. The branching of the spinal nerves to the back and to the ventral body wall (ramus dorsalis and ramus ventralis) occurs only as a secondary event following the innervation of the ectodermal ring.

The second tissue that becomes innervated in the embryo is the *musculature*; the basic type of innervation here is known as a *motor innervation*. Unlike the cells for sensory innervation that arise in the spinal ganglia, the cell bodies for motor innervation develop and remain within the densely packed ventral horn of the spinal cord (Fig. 4.10). *Motor neurites* grow out of the neural tube into the dilation fields of the muscles. Conversely, *sensory neurites* grow in the opposite direction from spinal ganglia into the neural tube. The same **spinal ganglion** cells are connected to the skin of the ectodermal ring by means of their dendrites. If we bear in mind that the membranes of these extensions, and especially the growing tips, are **permeable** to molecules, then we can expect that in a growing spinal cord, there is already a *reflex arc* that is closed by the metabolic movements between sensory and motor neurons. In all probability, the developing brain and spinal cord is already one powerful control system, for which the periphery has regulatory significance. Experiments on animal embryos probably could not prove this interpretation conclusively; however, they do support it.

How do the nerves find their way? In answering the last of our three questions, we return to the universal proposition: *Organs differentiate wherever there exists both a spatial opportunity and a metabolic occasion, that is, in a biodynamic field.* For the case of nerves, this proposition does not mean that developmental dynamics alone is sufficient to provide the nerves with a pathway, but rather, that the path-finding of nerves has developmental-dynamic properties. Therefore, the previous concept of nerves finding their innervation territories of their own accord, which ignores the role of developmental dynamics, is incorrect. In tissue culture experiments, it has never been possible to obtain a normal branching pattern of nerves. The pathway for nerve fibers is normally prescribed by the organs-to-be-innervated and is therefore laid down from without. We must assume that submicroscopic material (i.e., molecular) movements are decisive for this process; namely, that ordered metabolic movements work in a manner that determines the form of the incipient innervation pattern.

An analogy may help elucidate the argument. The course of a river cannot be explained on the basis of the development of its shipping, nor by a knowledge of its sources, its tributaries, or the specific locations of the harbors at its mouth. It is only the total topographical circumstances that determine the river's course. Similarly, the directions of growth flows (fluxes) within neurites or dendrites cannot be explained on the basis of their subsequent use or significance for human behavior patterns.

As is valid for all organs, including peripheral nerves, their design is a consequence of local differences in their growth. The metabolic movements and metabolism in the vicinity of growing dendrites are certainly quite different from that of growing neurites, otherwise they would not appear so different by light and electron microscopy. As a sign of their dissimilarity, one finds that dendrites and neurites grow into different regions of innervation. One therefore assumes that these regions contain metabolic fields with correspondingly different metabolic movements. We have stated above that peripheral dendrites invariably arise in connection with an epidermis that has thickened due to restriction of its surface growth, whereas neurites arise in connection with embryonic musculature that becomes slender as it is extended. It cannot be claimed, say, that muscles acquire their nerve supply because this serves to maintain the species, or because a nerve supply is useful for the execution of voluntary movements. Rather, muscles become innervated in the embryo because there exists both the spatial opportunity and the dynamic occasion for their innervation.

As an example, the head region of a 2.57 mm long human embryo is illustrated in Figure 4.13. With growth flexion of the head, the body wall (ectoderm plus inner tissue) in the vicinity of the flexion folds is restricted in surface growth and so thickens to form the facial part of the ectodermal ring. The tall cells over the flexion folds press against one another and so probably release some of their fluid contents. We have good grounds for believing that these fluids, which are released into the inner tissue beneath the ectodermal ring, contain substances (probably charged) that can be taken up by the growing dendrites. Thus, between the innervation area and the tips of the dendrites there must already exist material connections in the sense of pathways, well before the dendrites themselves have reached the skin. With this absorption of substances, the dendrites could elongate by apical appositional growth using the substances (catabolites) released by the ectoderm for synthesis (anabolism). By means of this material uptake, the dendrites are probably sucking themselves toward the tissue-to-be-innervated, and are therefore sustaining metabolic movements toward the cell body (**afferent**) and its nucleus (karyopetal). It is here proposed that apical growth with afferent metabolic movement is a characteristic of all growing dendrites.

According to this interpretation, the lips, the palms of the hands, the soles of the feet, and the digital pads are well supplied with nerves, not because these cutaneous regions must subsequently acquire specially differentiated capacities for touching and gripping, but because during embryonic development, the nerve endings are confronted by special growth stimuli on the part of a thickened epithelium (ectodermal ring).

Special prerequisites also exist for the growth of neurites. In order to illustrate this, a somite is depicted in Figure 4.15 during the growth dilation of its muscle fibers. During the growth elongation of the somite, a dilation field is arising immediately alongside the neural tube. This field is the segmental (metameric) anlage of a muscle. The field is represented spatially by a bundle of cells that are becoming slender. In each of the elongating muscle fibers, the cytoplasm constitutes a suction field that can draw in nerve endings. The possibility cannot be refuted that the suction field causes substances to flow from the terminals of neurites into the growing muscle cells. According to this concept, the nerve endings in muscle fibers only become motor nerves on account of their prior growth functions.

Naturally we cannot experiment on human embryos to test the above propositions on the growth dynamics of dendrites and neurites. Nevertheless, there is much evidence for (and none against) the views (i) that a developing dendrite sucks itself toward its innervation territory by material absorption and elongation of its apical region (growth cone), and (ii) that growing muscle fibers on their part exert suction forces and so draw neurites into the muscle field. These views compel us to assert that generally neural processes only become differentiated into dendrites and neurites, respectively, because they are subjected to different demands in growth. The following working hypothesis is therefore justified: In growing muscle cells, there exists a material need and therefore an occasion for suction, so that substances, and in particular ions, stream out of the ends of the motor neurites into the growing muscle fibers. In this way it is possible to understand, in a developmentally dynamic sense, why neurites have delicate fine tips initially but later terminate on muscle cells in broad contact zones (the so-called end-plates). Electron microscopical and physiological investigations of later stages have confirmed that so-called transmitter substances pass to the muscle cell from the motor end-plates. In any case, the so-called **efferent** nerve fibers prepare themselves, already as they are growing, for the subsequent efferent conduction of stimuli to the muscle. Conversely, the growth cones of dendrites are initially large, pale staining, metabolic factories whereas the tips of mature dendrites are especially delicate and pointed.

According to the above concepts, the difference between the pathfinding of **dendrites** and **neurites** lies in the dendrites sucking themselves of their own accord toward their source of supply, whereas vice versa, the neurites are being sucked by growing muscle fibers (Fig. 4.17). In either case, the direction of growth of the nerves appears to be programmed by the metabolic field of their environment and not, say, through genetic information or "encoding." What is certain is that the nerve processes that grow toward thickened skin subsequently conduct afferent stimuli and that those cell processes growing into muscle subsequently conduct efferent stimuli. There is no evidence to support the idea that sensory and motor nerve fibers are capable of finding their

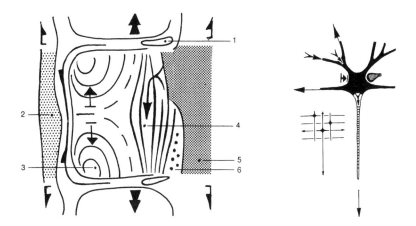

Fig. 4.15. Part of the neck region of a 4.2 mm long human embryo showing a somite in vertical section. Between ectoderm (lightly stippled) and the neural tube (densely stippled) lies a somite (refer to Figs. 2.31–2.34). The first motor neurites grow into the myotome (4). Thick black arrows and external half-headed arrows: growth elongation of the embryo. Arrows with cross-tails: growth expansion of the dermatome. Converging half-headed arrows: restraining function of the somitic capsule. Tailed arrow: growth innervation of the myotome. 1) dorsal spinal branch of aorta, 2) ectoderm, 3) dermatome, 4) myotome, 5) wall of neural tube, 6) sclerotome (dots).

Fig. 4.16. Schematic representation of a neuron. Cell body (soma) with dendrites black; neurite hatched. Ending of another nerve cell stippled. Large arrows at top: growth pull of dendrites. Tailed arrows: metabolic movements. Arrow with cross-tail: growth pressure of surrounding tissue bed. Long arrow at bottom: growth elongation of a neurite. Inset: spatial relation of cell bodies to the principal directions (trajectorial system) of their processes growing in interstices.

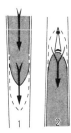

Fig. 4.17. Left: growth dynamics in a neurite; tailed arrows indicate growth with metabolic movements and release of substances. Right: growth dynamics in a dendrite; tailed arrow indicates appositional growth with uptake of substances; outlined arrow indicates direction of dendritic growth.

pathways of their own accord. Rather, these pathways seem to be pre-scribed in a topodynamic manner (*topos*, Greek; position). According the above argument, we propose the following working hypothesis: *the very growth of neuronal pathways is already an early nervous activity.*

Chapter 5

The Major Sense Organs
and the Face

We learn to recognize faces so efficiently: in a sea of strangers we can almost instantly identify the one familiar face. In this chapter, we will try to get to know the face of the human embryo and to answer rationally some questions about its development. The manifold faces of humankind, as well as the capacity of the face to change its expression, are portrayed for us by artists and actors. And yet who would anticipate that the high capacity for expression in the human countenance depends, among other things, on its earliest embryonic development.

A better understanding of the face can be obtained if we note the location of the major sensory organs at the time when they first appear. Eyes, ears, and nose are the typical elements for the construction of a face and these arise in very early embryonic stages, long before their higher functions become evident.

Ear

The *ear* seems to be one of the most complex organs. However, some simple aspects of the development of the ear can be highlighted if one starts by examining the ear's positional relations to neighboring structures.

Figure 5.1 shows a dorsal view of the upper half of a human embryo about 2 mm long, at the beginning of the 4th week of development. The neural groove is still open in the head region at the superior neuropore. Here, the distance between the two neural folds in the vicinity of the flexed head is greater than it was previously in the region of the neck. The appearance is reminiscent of a longitudinal skin wound on the back of a finger that splits open as the finger is flexed. As the neural groove gapes open, the dorsal neural folds tilt markedly to the side and, on each side, in a region indicated in Figure 5.1 by converging arrows, the surface growth of the ectoderm becomes restricted. In this zone, the ectoderm thickens locally and forms a so-called placode (**otic placode**) composed of tall ectodermal cells. The otic placode then becomes a depression, and then a vesicle. This invagination process occurs because cell division occurs mainly at the exposed surface of the placode adjacent to the (amniotic) fluid, whereas cell growth occurs mainly at the base of the thickened ectoderm, adjacent to the nutrient-rich stroma. The greater growth at the base enables the placode to arch inwards, first to form a pit and then by closing, a vesicle, the **otic vesicle** or **otocyst** (Fig. 5.2). The thinner ectoderm around the placode continues to

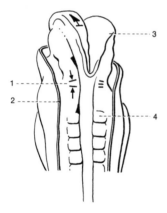

Fig. 5.1. Head and neck region of a human embryo about 2 mm long. Dorsal view. Upper arrow with cross-tail: longitudinal growth and bending of the still open neural tube. Converging half-headed arrows: restraining function of the left dorsal aorta. Converging arrows: restricted surface growth in the region of the anlage of the otic placode. 1) ectoderm, 2) cut edge of the amnion, 3) transition zone between ectoderm and neural epithelium (neural crest), 4) somite.

increase in surface area until it finally closes over the vesicle. During this period, many cells are being squeezed out of the wall of the otic pit and vesicle to become inner tissue cells (the so-called **otic neural crest** or **otic mesectoderm**) along with cells from the crest of the nearby neural fold.

We already know from Chapter 4 that metabolic movements occur regularly in the metabolic field of a thickened ectoderm, and that these molecular movements enable contacts to form between the ectoderm and dendrites. As soon as the otic placode has a sufficient number of cells, dendrites from the otic neural crest and the margin of the nearby neural tube (hindbrain) grow appositionally toward the placode. Thus the placode is said to receive a sensory innervation. The ingrowing dendrites to the otic vesicle represent the *vestibulocochlear nerve*. So long as the adjacent ectoderm stays thin, it will remain uninnervated.

When seen from the side, the otic vesicle of a 10 mm long embryo has the outline of a comma (Fig. 7.19). The dorsal blunt end of the

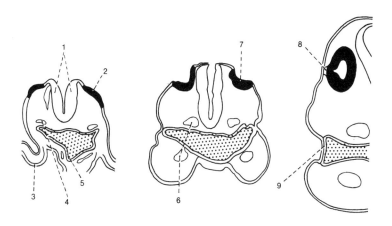

Fig. 5.2. Three stages in the development of the ear anlage (black) of 2 mm, 2.6 mm, and 4.2 mm long human embryos. Left: stage of otic placode (2). Middle: stage of otic pit (7). Right: stage of otic vesicle (8). 1) neural groove, 3) rim of umbilicus, 4) and 6) aorta, 5) foregut endoderm (lumen stippled), 9) ectoderm of 1st pharyngeal groove in close contact with endoderm of 1st pharyngeal pouch (anlage of tympanic membrane). (after Arey, 1946).

comma is the anlage of the membranous **labyrinth** with the *endolymphatic sac* on its medial aspect, and the ventral narrower end is the anlage of the *cochlea duct* (containing the organ of hearing). As the otic vesicle grows, the increase in surface area at the blunt end outstrips the increase in volume. The dorsal part of the vesicle collapses. A positional feature that may be associated with the failure of the volume of the blunt part of the otic vesicle to keep pace with its surface growth is the proximity of the endolymphatic sac to the developing sigmoid sinus in the meninx of the brain: it is likely that there is a continual leakage of fluid from the more "stagnant" otocyst through its endolymphatic sac into the nearby vein. In any event, parts of the walls of the labyrinth gradually grow into contact with each other. At these sites, according to the rules applying to corrosion fields (Chapter 3), the ectodermal cells die away, leaving behind only the peripheral zones of the otic vesicle to continue further development. In this way, the three so-called *semicircular ducts* arise at the perimeter of the wide part of the otocyst. There is an anterior, a lateral, and a posterior semicircular duct. Already in embryonic stages, as the head changes position, fluid flows along the walls of the three mutually perpendicular ducts (similar to the movement of a liquid in a spirit-level) thereby leading to stimulation of the sensory dendrites. These early functions are a precondition for the subsequent ability of the labyrinth to function as an organ for the maintenance of body equilibrium.

At its ventromedial edge (i.e., the sharp end of the comma), the otic vesicle is innervated by the *cochlear nerve* (dendrites of neural crest cells). The nerve accompanies this small portion of the otic vesicle as its epithelium grows in surface area and the tip of the comma elongates. However, in comparison to the epithelial surface growth, the nerve lags a little behind. Due to the restraining action of the nerve fibers described in Chapter 4, the tip of the otic vesicle rolls inwards as it grows, so forming the cochlear duct of the inner ear, which spirals around the central axis of the cochlea (*modiolus*) containing ganglion cell somas and their cytoplasmic extensions.

As the side of the head in the vicinity of the visceral arches tilts laterally with growth, the first two (large) visceral arches become compressed dorsoventrally and acquire transverse creases. The transverse creases are compression folds caused by the dorsolateral growth of the brain above and the ever enlarging, beating heart below. These creases constitute the initial, complex relief-formation of the external ear, the anlage of the *auricle*. The intricate folding of the external ear is quite striking even in adults. The young auricle is initially just a partial relief of the lateral neck region (Fig. 5.3). Isolated bumps or so-called auditory hillocks that are not related to any of the above dynamic factors are never observed in young human embryos. The groove between the 1st and 2nd visceral arches is very deep and the foregut endoderm of the 1st pharyngeal pouch lies so close to the ectoderm that the ear-drum (*tympanic membrane*) arises here as a thin sheet of tissue. As long as the face remains relatively small, the growth of the brain forces the side of the head to incline obliquely outwards. The tympanic membrane also conforms to this oblique inclination, just like the auricle.

The tympanic membrane, being the floor of the groove between the 1st and 2nd visceral arches, is the boundary between the **external auditory meatus** and an internal passage-way, which is the *auditory tube* (of Eustachio) associated with the 1st pharyngeal pouch. This tube remains throughout life as a deep niche in the lateral wall of the pharynx. When we catch a cold, the walls of this tube can swell and its closure can temporarily impair our ability to hear.

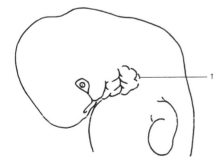

Fig. 5.3. Head region of an 11 mm long human embryo already showing a distinct anlage of the auricle (folds in the face–neck transition region). 1) anlage of auricle.

The blind-ended part of the auditory tube (deep to the tympanic membrane) is called the *middle ear*. The inner tissue here is covered by endoderm and is derived from the 1st and 2nd arches. Toward the end of the embryonic period, tiny folds arise in the endodermal walls of the middle ear. Densation fields arise within the stroma of these endodermal folds, with the shape of the fold determining the shape of the underlying densation field. The densation fields become contusion fields of chondrocytes and, following the piston-like growth swelling of cartilage, eventually become extensively ossified. The foci of ossification are the anlagen of the *auditory bones* (ossicles): the hammer (malleus), the anvil (incus), and the stirrup (stapes). The auditory bones are covered by a thin layer of endoderm and remain attached by ligaments to the endodermal walls from which the folds originally arose. The ossicles are the first bones to attain their adult shape and size; they are almost fully ossified at birth.

Eye

The famous physiologist Helmholtz once commented that it would be hardly possible to find a poorer optical apparatus than the human *eye*. The lens of the eye is made from cellular fibers and is so inhomogeneous that an optician would scarcely want to use it. The light-sensitive layer of the eye, containing the photoreceptors, does not lie next to the glassy (vitreous) body that transmits the light, but behind layers of cells and fibers that are actually in the way of the incoming light. Therefore, from a photographic point of view, the light-sensitive layer lies on the wrong side of the retina. Furthermore, before they can even penetrate the retina, the light rays have to first pass a network of blood vessels at the inner surface of the ocular fundus facing the vitreous body.

Once again a glance at the ontogeny of the eye will help us to comprehend this puzzling anatomy. The eye anlagen arise early at either side of the cranial end of the neural groove in embryos about 2 mm long, well before the superior neuropore has closed. Initially, a groove called the *optic sulcus* arises in a transverse arc near the superior end of the neural groove (Figs. 5.4, 5.5). Seen from the front, the optic sulcus appears as a local, outward buckling of the left and right walls of the neural groove. The floor of the optic sulcus grows close to the ectoderm and is also anchored laterally in the adjacent inner tissue. As the superior neuropore closes, the sulcus can no longer be seen from outside

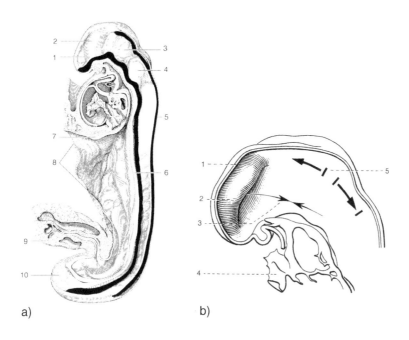

a)   b)

Fig. 5.4. a) Human embryo about 2 mm long (Carnegie Stage 10) sectioned in median plane; overview for (b). The lumen of the neural tube lies between the two thick black lines; the tube is still open above and below (superior, inferior neuropore). 1) entrance to the right optic vesicle (optic sulcus). 2, 3, 4) lateral parts of the brain wall, 5) dorsal wall of the neural tube, 6) notochord, 7) anlage of liver, 8) cut edge of the wall of yolk sac (yolk sac duct wall), 9) body stalk, 10) undifferentiated tissue of end bud (former primitive streak region). b) Head region of same 2 mm long embryo shown in (a) viewed from left. The ectoderm and underlying stroma have been removed revealing the left half of the neural tube and the anlage of the left optic vesicle (entrance to optic vesicle lies beneath plane of diagram). 1) cut edge of the surface ectoderm, 2) anlage of optic vesicle (wall adjacent to surface ectoderm), 3) restraining function of the inner tissue, 4) stem of aorta near heart (truncus arteriosus), 5) growth elongation of the neural tube.

the embryo. On account of neuropore closure, each eye anlage has now developed into a cul-de-sac or **diverticulum** of the lumen of the forebrain, the so-called *optic vesicle* (Fig. 5.6). The vesicle lies immediately beneath the ectoderm. With increasing surface growth of the wall of the optic vesicle, its lumen (optic ventricle) narrows. As the optic ventricle narrows to a slit, the wall of the optic vesicle adjacent to the ectoderm approaches the less superficial wall; the former invaginates, as it were, into the lumen of the optic vesicle.

As a consequence of this invagination, the optic vesicle is now called the *optic cup*. Henceforth, the cup consists of two walls lying close together and is therefore bilaminar. The inner wall facing the surface ectoderm becomes the *retina* while the outer wall becomes the *pigment epithelium*. The rim that folds over between the two walls forms the *iris*

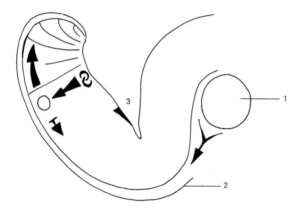

Fig. 5.5. Drawing of a horizontal hemisection through the developing eye of embryo, about 2 mm long; midline is at top; superior neuropore is at top left. Broad straight black arrow: displacement (not "migration") of daughter cells toward the outer surface of the neural epithelium following cell division in the ventricular zone. Diverging double arrows: surface growth, which, together with a similar contralateral growth, leads to closure of the neural tube in the region of the forebrain. Half-headed arrow near 3: a site of little growth (growth pull) at the inner surface of the eye anlage. Tailed arrow: direction of nutrient movement in stroma. 1) blood vessel, 2) ectoderm covering the optic vesicle, 3) lumen of optic vesicle leading into the lumen of the neural tube.

with the *pupil*. The pigment epithelium is the first black-pigmented tissue to appear in the human embryo. The cells of this epithelium support the nourishment of the retina, to the extent that they furnish the retina with a major part of the nutrition that they themselves have absorbed from the adjacent stroma. The unique feature of these metabolic movements lies in the fact that the one epithelium is withdrawing nutrients from the underlying stroma as well as maintaining a nutritional supply to a neighboring epithelium. The significance of the connections between retina and pigment epithelium is still evident in adults, where a detachment of the retina from the pigment epithelium leads to blindness.

The optic cup, which has developed from the optic vesicle, has the form of a tiny clasping hand by means of which the brain is able, as it were, to take hold of a piece of skin as the anlage of the lens. At first the *lens* is also an ectodermal placode, then an ectodermal pit, and finally an ectodermal vesicle. It is only when the wall of this vesicle thickens by means of an intensive uptake of nutrients and simultaneously releases the watery by-products of metabolism into the superficial, adjacent stroma (thereby forming the anterior chamber of the eye) that the vesicle becomes the crystal-clear lens that later serves to regulate the refraction of light.

At this point, it is worth reminding ourselves that the optic vesicle is part of the neural tube, joined to the latter by a slightly narrower tube

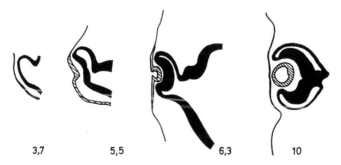

<table>
<tr><td>3,7</td><td>5,5</td><td>6,3</td><td>10</td></tr>
</table>

Fig. 5.6. Development of the optic vesicle into the double-walled optic cup (black). Importance of the optic cup for the development of the lens: the optic cup "grasps" the ectoderm. Lens cross-hatched.

called the *optic stalk*. The cells of the optic vesicle wall are wedge-shaped and are aligned radially in a direction toward the nourishing inner tissue, that is, toward the periphery of the vesicle. As the optic vesicle transforms into the optic cup, the cell processes (extensions) that were originally in the vicinity of the overlying skin now lie on the inner wall of the cup directed toward the lens. As the initially very small optic cup grows larger, its walls begin to flatten. The continual flattening results in the processes of those cells in the region of the anlage of the ocular **fundus** becoming extended at right angles to the radial processes of deeper cells. In this way, an ordered layer of optic nerve fibers arises across the inner surface of the retina. Thus it is possible to comprehend how it happens that light must first traverse a nerve fiber layer at the inner surface of the retina before reaching the photoreceptor layer. The nerve fibers grow appositionally out of the retina along interstices in the optic stalk toward other parts of the nervous system.

Nose

The *nose* also makes its first appearance in the embryo as a pair of placodes. Such placodes always signify a region of restricted surface growth: at the time that the *nasal placode* is appearing, a narrower sheaf or strand of inner tissue is found at its base. As it passes dorsally over the optic stalk, this strand of tissue is sandwiched between the adjacent walls of the neural tube and the optic vesicle; the strand is impeded in growth and so becomes a restraining apparatus (retension field of the nasal **gubernaculum**). Where this compressed inner tissue radiates toward the skin, it restricts the surface growth of the overlying ectoderm (Fig. 5.7). This restriction of surface growth leads to a local thickening of the epithelium in the form of the nasal placode mentioned above. The placode soon transforms into a pit of thickened epithelium, so that the anlage of the *nostril* appears already at an embryonic stage. As there are two eyes, there are also two nasal placodes. Initially both nasal placodes develop laterally, corresponding to the wide separation between the eyes. It can be said that the nose is a developmental product of the eye and the brain. At the beginning of the 2nd month, the human nose

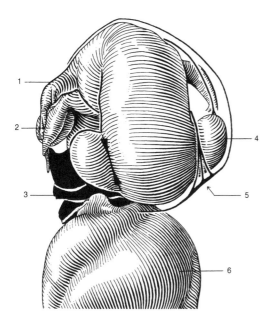

Fig. 5.7. Part of a reconstruction of the head region of a 3.4 mm long human embryo. Endoderm black. Converging double arrow: restraining function of the inner tissue strand between the optic vesicle and the forebrain. The nasal placode arises at the site where this strand merges with the ecto-derm (5). 1 and 2) cranial nerves, 3) 2nd pharyngeal pouch, 4) optic vesicle, 5) nasal placode (olfactory groove), 6) pericardium. Refer to the earlier stage depicted in Fig. 2.36.

appears wider than it is long. The elongation of the nose does not occur until the whole face starts to lengthen.

The Broad Face of the Embryo

Even in an embryo 16.2 mm long in the 7th week of development (Fig. 5.8), the face is confined between the growing brain and the power-ful heart bulge; the face is therefore broad. The distance between the eyes, as well between the nostrils, is still relatively wide. The cleft of the mouth is oriented transversely. Compressed between the forward-arching forehead and the nose, the inner tissue between the eyes is

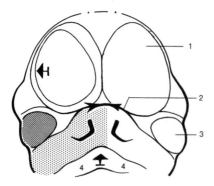

Fig. 5.8. Forehead and face region of a 16.2 mm long human embryo. Development of the forward direction of gaze. Nose–cheek region stippled light gray. Eye dark-gray. Lower arrow: direction of growth pressure from the heart against the head during the increasing flexion of the embryo. Converging double arrow: restraining function of the stroma between the nose and the forehead. Upper arrow: transverse widening of the cerebral hemisphere above the eyes. 1) cerebral hemispheres, 2) tense connective tissue between the left and right eyelids (interorbital ligament), 3) eye, 4) first flexion fold (mandible).

restricted in growth and becomes stretched transversely between the eyelids of either side. This stretched band of tissue is known as the *interorbital ligament*; it forms, so-to-speak, the embryo's spectacle-bridge. For a long period, this ligament anchors the medial region of each eye in its embryonic position as the back of the head broadens with the formation of the temporal lobes of the brain. Thus, the distance between the eyes hardly increases and the gaze appears to be directed gradually more to the front. Therefore, we can say that the typical frontal gaze of the human is a direct consequence of the development of the brain. Animals with a smaller brain and a smaller heart have a different positioning of the eyes from that found in humans.

Long before developmental movements were recognized, physiognomists realized that human expressions could be interpreted psychologically as the result of interplay between the activities of the brain and the heart. This corresponds to the experience that, on the one hand, the rapid reactions of personal behavior that manifest themselves in facial mimicry may be related to striking changes in heart activity and that, on the other hand, long reflection frequently leads to somatic sensations in the ocular and frontal regions of the head, or even to headache. We usually separate the feelings of the heart from those of the head.

Organs in the Region of the Foregut

Toward the end of the 2nd month, when the embryo is about 30 mm long, the distance between the heart and brain begins to increase. The face therefore begins to elongate.

Lips: In an embryo about 30 mm long, whose face is just starting to elongate, the *facial skeleton* consists of the skeletal anlage of the upper jaw and nose (nasomaxillary skeleton) on the one hand, and the cartilaginous skeleton of the lower jaw (mandible) on the other. Both foci of skeletonization, as viewed from the side, make a forward opening angle between each other. As the diverging sides of this angle enlarge with the growth elongation of the face, the embryonic skeleton of the mouth also enlarges and the ring of inner tissue around the opening of the mouth becomes stretched. The *circular musculature of the mouth* (orbicularis oris muscle) arises in this dilation field, its development conforming to the general rules of a dilation field. With increasing growth extension of the circular musculature, its resistance to further stretching also gradually increases. This causes the margins of the mouth, namely the **lips**, to roll inwards and so the mouth (that is, its soft part) closes externally (Fig. 5.10). Behind the lips, however, the mouth cavity is broadening in all directions and thus increasing in volume. A suction zone therefore arises; the embryo as it were "suckles." The so-called suckling reflex of the newborn child is a late consequence of this early developmental act and not, say, a recapitulation or an atavism of early phases of phylogeny. Already in the 2nd month, the development of the nervous system is taking part in this process.

Teeth: As soon as the lips of the embryo start to roll inwards, the mucous membrane at the margin of the lips becomes compressed against the deeper mucous membrane lining the mouth, and therefore is hindered in its surface growth (Fig. 5.10). As a consequence of this restriction in surface growth, the epithelium thickens and, coupled with this, the adjacent inner tissue condenses in an orderly manner along the base of the lips. A molding of ectoderm and inner tissue known as the *dental lamina* arises along the upper and lower jaw-line, respectively. The *tooth germs* (Fig. 5.11) originate from this dental lamina, according to precisely known kinetic rules. Over months of development, the tooth germs differentiate into *teeth* that grow more and more toward the mouth cavity. After birth the first teeth, known as milk teeth, break through the mucous membrane. The definitive set of teeth appears later. The foregoing developmental movements of the young tooth germs are an early factor in the act of biting. In other words: the formation of the teeth is an embryonic performance that represents growth biting. Growth biting is a prerequisite for subsequent biting. The development of teeth is another example of a functional development, that is, the development of a pattern of behavior through growth.

Tongue: In a similar manner, the emergence of the human tongue chronicles preliminary functions for speaking. The fundamental basis for the well-known, great mobility of the tongue lies in its own development. In this respect the following developmental movements can be demonstrated. As the growing embryo bends forward, the foregut becomes curved. In a 7 mm long embryo, the surface growth of the wall of the foregut adjacent to the brain is more favored than the surface growth of the wall that constitutes the floor of the mouth. The roof becomes thinner and the floor of the oral cavity, relatively hindered in its surface growth, becomes thickened. The thickened epithelium is the epithelial anlage of the *tongue*. Then, as the mouth cavity increases in size, the tongue anlage also grows in proportion (Figs. 5.12, 5.13). Initially the tongue has the form and the size of one of the many tiny papillae that one can observe later near the root of the adult tongue. This corroborates the thesis of the similarity of the parts to the whole (encapsis), as can be observed in the most divergent of organs. The increasing forward arching of the tongue epithelium precedes that of the underlying inner tissue. Within the tongue, the inner tissue cells become stretched lengthwise (longitudinally) during the growth elongation of the entire anlage of the tongue. Afterwards, younger inner tissue cells become stretched transversely to the long axis of the tongue as the foregut, and thus the tongue, broadens with head growth. Still later, other cells will gradually become dilated in a direction perpendicular to each of the above fields (i.e., vertically). Collectively, these dilated cells constitute the trelliswork of the intrinsic **lingual musculature**.

Because of the multitude of muscle fibers in this three-dimensional network, the lingual musculature becomes extremely well innervated at a very early stage. The tongue thus becomes associated with a special functional center in the brain. The capacity of the lingual musculature for such a high degree of cerebrally directed activity is a characteristic feature of a uniquely human cerebralization. Only humans are capable of highly differentiated speech. The very development of the tongue is a prerequisite for, and a prelude to, the subsequent act of speaking. Speech is prepared for phenogenetically by the functional development of the tongue. The process of learning to speak only completes a performance that early development has, to a very great degree, already initiated.

The Palatal Processes: If we artificially open the mouth of an embryo about six weeks old, by anatomical dissection, we find that the roof of the mouth cavity contains two very thin regions. Each thin region in the roof is opposite the floor of the adjacent nasal pit. The nasal ectoderm and the endoderm of the roof of the mouth lie so close to each other here that a corrosion field normally arises. The location of the two internal nostrils (**choanae**) is determined by these corrosion fields. As the cellular tissue disintegrates in these fields, the two former nasal pits become short nasal passages that open directly into the mouth cavity.

In the early days of the 2nd month, the lumen of the foregut is still not divided by an intermediate roof. However, we can identify the anlagen of this subdivision arising on the lateral walls of the foregut as a thickening of the floor of each eye socket (orbit, Fig. 5.9). Along the free internal border, the epithelium arches convexly into the mouth and is therefore a diverging wedge epithelium covering a crest; the crest is the precursor of the so-called **palatal process**. Corresponding to the divergence of its cells, this wedge epithelium grows farther into

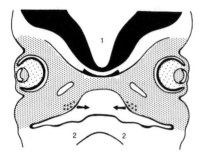

Fig. 5.9. Section of the 16.2 mm embryo in a frontal plane through the eyes and the palate region. Brain black. Converging half-headed arrows: restraining function of the dura. The two small converging arrows indicate the initial growth movement of the developing palatal processes in the growth region under the eyes. Dots in connective tissue adjacent to these arrows indicate stretched connective tissue cut in cross-section (nasomaxillary ligament). 1) brain ventricle, 2) region of lower jaw.

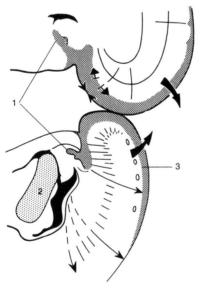

Fig. 5.10. Growth movements of the lips. Converging double arrow: slight surface growth of ectoderm. Diverging tailed arrows: more rapid surface growth. Simple solid arrows: in-rolling movement of the lips. 1) dental lamina with tooth germs, 2) cartilaginous lower jaw (mandible), 3) interface between thick ectoderm (stipple) and underlying inner tissue with blood vessels (ellipses). Curved arrows in lower jaw: tension lines in inner tissue. Bone at the edge of the cartilaginous mandible black.

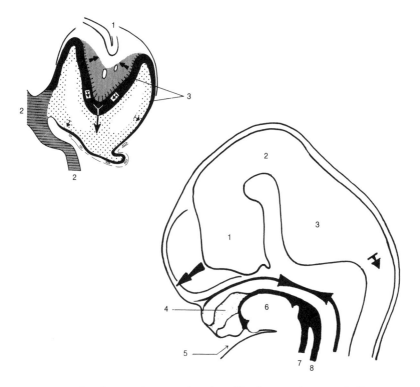

Fig. 5.11. Tooth germ in upper jaw (maxilla; from region near indicator 1 in maxilla of Fig. 5.10, but at a later stage of development). 1) connective tissue papilla with vascular loops (anlage of tooth pulp), 2) dental lamina, 3) inner and outer enamel epithelium (ameloblasts) black. As the inner layer of ameloblasts forms enamel, the outer layer becomes stretched and disintegrates. Arrows with cross-tails: growth pressure. Tailed arrow: release of fluid into bell-shaped enamel organ. Solid arrows: growth movement of inner enamel epithelium.

Fig. 5.12. Right half of the head region in a 17.5 mm long embryo. Forming function of the stretched connective tissue of the cranial base (basicranium, converging double arrow). The single arrows indicate the growth expansion of the brain forwards and backwards over the base of the skull. Foregut lumen (black) is arched under the skull-base, which is still composed of stretched connective tissue. 1) interbrain (diencephalon), 2) midbrain (mesencephalon), 3) hindbrain (rhombencephalon), 4) right nasal passage, 5) entrance to mouth, 6) tongue, 7) trachea, 8) esophagus. Based on a serial section reconstruction.

the lumen of the foregut. The palatal processes unite in the midline and so divide the initially single foregut lumen into two stories. During the 2nd month, the top story or nasal cavity is further subdivided into right and left chambers by the growth of the *nasal septum*. Sometimes the gap between the palatal processes (*palatal fissure*) may fail to close and the resulting anomalous aperture (cleft palate) may become so large that it causes disturbances in breathing and speaking.

The Long Face: As soon as a densation field arises inside the bridge of the nose and inside the nasal septum, and cartilage develops in this field, then the young (discoidal) cartilage cells gain the capacity, by their growth swelling, to exert a piston-like action mainly in the direction of the tip of the nose. The growth of this cartilage causes the nose to project externally, and even more so, as the entire face is now elongating due to the growth pull of the diaphragm as the powerful embryonic heart and the intestines move farther away from the head (**ascensus** and **descensus**, Fig. 5.14; also see Chapter 7). The face now obtains extra space for surface growth between the ascending brain and the descending heart; it develops into the typically long human face.

In the developmental period of facial elongation, the connective tissue deep within the face forms into a kind of muzzle stretched over the growing expansive *cartilaginous skeleton of the nose*. This "nasal muzzle" is attached above to the cartilaginous **basicranium**, and below to the compressed tissue of the lower jaw and the hyoid arch (Fig. 5.15). As the muzzle becomes more stretched, so the nasal skeleton enclosed within it becomes more and more oblong. The mucous membrane on the inner aspect of the lateral wall of the nasal skeleton buckles and folds into the nasal cavity (Figs. 5.13, 5.15). The densation fields of the skeleton of the **nasal conchae** arise within these folds after they have grown sufficiently; this is another example of outside–inside differentiation.

Once again, all the growth functions described above are a preparation for later performances. Whatever has not already been practiced before birth cannot be further elaborated after birth and then little-by-little put into effect. The whole of early development is a growth function that is an absolute prerequisite for subsequent behavior. Everything that we later call activity starts with growth functions.

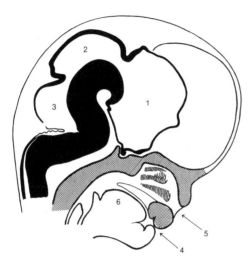

Fig. 5.13. Left half of head region of a 24 mm long human embryo. Commencement of the separation of the head region of the gut into two stories. The triangular light zone above the tongue (6) is the left palatal process growing out from the lateral wall of the pharynx. The cut surface of the skull-base and the upper jaw with upper lip are stippled. The three left nasal conchae are cross-hatched. 1) interbrain (diencephalon) between the two cerebral hemispheres, 2) midbrain (mesencephalon), 3) cerebellum, 4) entrance to mouth, 5) nostril, 6) tongue. (Based on a dissection of a preparation fixed in alcohol).

Fig. 5.14. Human fetus 37 mm long, in 3rd month of development at the time of elongation of the face. Arrow: descent of the viscera.

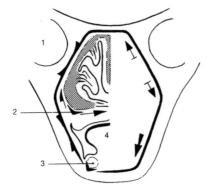

Fig. 5.15. Frontal section through the facial region of 37 mm long fetus (Fig. 5.14). Formation of the long face with stretching of the connective tissue of the nasal capsule (the "muzzle"). Converging half-headed arrows: restraining function of stretched connective tissue. Diverging arrows: piston-like growth and growth expansion of the nasal cartilage (stippled). 1) eye, 2) focus of bone in maxilla, 3) cartilage of lower jaw with external bony focus (black), 4) tongue. Two conchae are seen in cross-section on the lateral wall of the nasal cavity.

Chapter 6

The Organs of Movement

For almost 2,000 years, muscles have been interpreted as the active agents of body movements and the skeleton, as the passive supporting component. This is certainly true for the later stages of human development. However, as we know that the performances of adults are based on early growth functions and that each function is preceded by a unique development of function, it suggests that the skeleton and musculature should be investigated from the viewpoint of their early functions. These investigations have led to the completely unexpected finding that all muscles arise in dilation fields, that is, muscles initially function passively under stretch. Before muscles can contract, they must first be dilated. In contrast to the musculature, the developing cartilaginous skeleton functions initially as the active part of the movement apparatus. As we know (Chapter 3), the growth swelling of cartilage represents a piston-like function of the embryonic movement apparatus.

Axial Skeleton

The adult *skeleton* consists of the skeletal components of the head, trunk, and limbs; the *axial skeleton* includes the skull, hyoid bone, vertebral column, ribs, and sternum. If we abolish all preconceptions, then at first it is impossible to make sense of the form of the axial skeleton as it is usually represented (e.g., Fig. 6.1). Concerning even the position of the axial skeleton, it is initially difficult to comprehend why, say, the vertebral bodies are located ventral to the spinal cord and

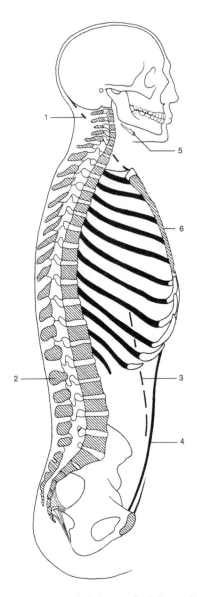

Fig. 6.1. Axial skeleton of adult. 1) oblique lateral neck muscle (sternocleido-mastoid muscle), 2) spinous process on dorsal side of 1st lumbar vertebra, 3) oblique abdominal muscle (external abdominal oblique muscle), 4) straight abdominal muscle (rectus abdominis muscle), 5) hyoid bone, 6) sternum (ribs black).

not dorsal. We could ask further: Why do ribs only appear in the wall of the thorax and not in the wall of the abdomen? Or: Why is the base of the skull first made of cartilage whereas the roof of the skull transforms directly into bone? However, if we study the process of formation of the axial skeleton (i.e., the nature and manner of its skeletonization), then the rules of its positional development (topogenesis), formal development (morphogenesis), and structural development (tectogenesis) are revealed.

Therefore, once again, let us ask the embryo to reveal the ontogeny of these structures! In Figure 6.2, the facial flexion folds and lateral body wall of a 4.2 mm long human embryo are emphasized by stippling. This stippled zone continues over the region of both the face and the future external genitalia to the other side of the embryo. Taken as a whole, the stippled zone indicates a ring-like segment of the body wall where the ectoderm is thickened, the *ectodermal ring*. In classical comparative anatomy, the lateral sections of this ring were misinterpreted as milk-lines, despite the fact that they also occur in the embryos of reptiles.

In human embryology, we now know the following facts about the ectodermal ring. In the region of the ring, the skin is restricted in its surface growth and so thickens. Transverse sections through embryos of appropriate ages show clearly that the rapid dorsal growth of the spinal cord stretches the skin in its immediate neighborhood. Similarly, at the

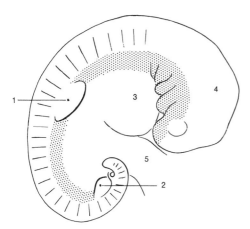

Fig. 6.2. Human embryo 4.2 mm long, age 28 days (Carnegie Stage 13). Ectodermal ring (thickened skin on lateral body wall) stippled. 1) anlage of upper limb, 2) anlage of lower limb, 3) heart bulge, 4) hindbrain, 5) umbilical cord.

ventral side of the embryo, the rapid growth of the heart–liver mass stretches, and thus thins, its adjacent body wall. Conversely, the body wall between these two rapidly expanding zones becomes compressed into a narrow strip. Thus the epithelium here cannot spread out and so thickens; the underlying stroma becomes compact. In the depths of the stroma under this ring, a field of tissue thickening (densation field) arises adjacent to the notochord. This densation field extends continuously from one side of the embryo to the other (Fig. 6.3). The crossover region lies ventral to the spinal cord and dorsal to the viscera of the trunk; it is the site of development of the axial skeleton, which constitutes the core of the entire subsequent movement apparatus. This densation field encloses the *notochord*, which therefore subsequently traverses the *vertebral column*.

In Chapter 3, we gave a survey of metabolic fields and described the development of a contusion field associated with the embryonic vertebra as an example of the differentiation of cartilage. Detailed anatomical investigations of human embryos of different ages have provided evidence that the neural tube (anlage of spinal cord) is the structure that initiates this contusion field, and that it does so in the following manner. Dorsally over the spinal cord, the wall of the vertebral canal is thin and therefore the growth resistance of this wall, in opposition to the growth of the spinal cord itself, is weak. Conversely, ventral to the spinal cord, the wall of the vertebral canal is thick (Fig. 6.4). As the spinal cord grows, above all in a dorsal (eccentric) direction against the skin, the **dura mater** flattens ventrally; the strengthening of the dura occurs initially only in this same ventral region. The ventral flattening of the dura represents a decrease in its curvature, that is, a partial straightening. This leads to the formation of a densation field immediately adjacent to the convex (outer) side of the thick ventral dura. In this field, the connective tissue thickens and so becomes the anlage of the *vertebral column*. Subsequent ventral flattening of the dura transforms the densation field into a contusion field (Fig. 3.21). There is no evidence to support the concept that cells from the somite (sclerotomal cells) "migrate" actively to the vicinity of the notochord to form the axial skeleton; the densation field around the notochord simply arises *in situ* as the remainder of the somite is dragged dorsolaterally by the growth of the neural tube.

The densation field where the vertebral column arises extends laterally into each of the interstices between the segmental arteries, which arise in a craniocaudal sequence from the dorsal aortae. Thus, each *vertebral body* is able to enlarge by means of these lateral extensions. The

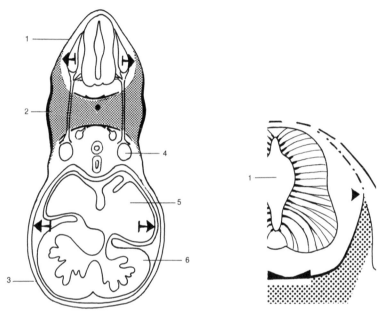

Fig. 6.3. Cross-section of human embryo 8.1 mm long (viewed from inferior aspect) showing thick and thin regions of body wall. Diverging arrows with cross-tails: growth expansion. Stipple: anlage of movement apparatus. 1) thin skin over rapidly growing neural tube, 2) ectodermal ring (thick skin in region of restricted surface growth), 3) thin skin over rapidly growing heart, 4) vein (superior cardinal vein), 5) right atrium of heart, 6) right ventricle.

Fig. 6.4. Vertebral canal and spinal cord of 11 mm long human embryo. Dot–dash line: thin part of dura over eccentrically growing spinal cord. Converging arrows: restraining function of dense meninx (dura mater), also perichondrium of cartilage. Arrowhead: fluid pressure of intercellular substance in endomeninx (arachnoidea); this fluid pressure has led to the formation of a contusion field in the vicinity of the vertebral column. Cartilage stippled. 1) spinal canal.

arrangement of aortic dorsal branches, which are tethered to the spinal cord, determines the segmental subdivision of the skeleton ventral to the spinal cord (Fig. 6.5).

What is valid for the vertebrae also applies to the *ribs*. Ribs too arise through cell condensations (densation fields) forming in the interstices between the craniocaudal sequence of thoracic blood vessels that arise as lateral branches of the segmental aortic rami. Kinetically speaking, the following description is an overview of what happens. Due to increasing enlargement of the heart–liver mass, the stroma adjacent to the convex side of the pleura (parietal pleura) becomes flatter, its curvature diminishing. In this zone of flattening, between the segmental vessels, tissue condensations arise that develop into precartilage segments, the anlagen of the ribs. The ribs then elongate by piston-like growth of disc-shaped chondrocytes; most ribs (true ribs) grow to join

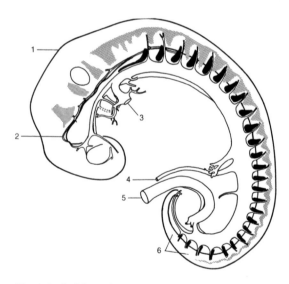

Fig. 6.5. Serial-section reconstruction of 4.2 mm long human embryo (Carnegie Stage 13). Anlagen of lateral processes of vertebral column, black. These processes arise in vessel-free zones between the evenly spaced segmental (metameric) arteries branching from the dorsal aorta (shown white). Neural crest and nerves, gray stipple. 1) ectoderm, 2) artery, 3) root of aorta (truncus arteriosus), 4) yolk sac (vitelline) artery, 5) umbilical artery, 6) spinal cord.

the breastbone (sternum). However, adjacent to the ventral surface of the liver, where its growth is especially intensive, the ribs never gain the opportunity to unite with the sternum (false and floating ribs).

Skull

As shown previously (Fig. 2.12), the 0.23 mm long endocyst disc possesses a broad upper end and a narrow lower end. The blunt end is the head region, and the pointed end is the trunk. The head region occupies more than half the length of the whole endocyst disc. Even a 7 mm long human embryo is mostly head. And even a newborn child still has a relatively small trunk.

As before, the differentiation of the *skull* cannot be comprehended in isolation, but only in relation to the development of its neighboring organs, especially the brain. The growing brain is anchored to the region of the face by means of blood vessels and nerves. As the brain bends more and more at its crown in the vicinity of the midbrain, additional flexures appear in the region of the forebrain and hindbrain (Fig. 6.6). We designate the subdivisions from superior to inferior as follows: endbrain (*telencephalon*), interbrain (*diencephalon*), midbrain

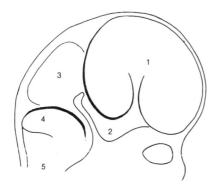

Fig. 6.6. Schematic view of brain of embryo, about 28 mm long. 1) cerebral hemisphere (telencephalon), 2) interbrain (diencephalon), 3) midbrain (mesencephalon), 4) anlage of cerebellum (rhombic lip), 5) junction of hindbrain and spinal cord (medulla oblongata).

(*mesencephalon*), and hindbrain (*rhombencephalon*) with *cerebellum* ("little cerebrum"). Corresponding to their early development, the interbrain and hindbrain have extensive connections via nerve fibers with the eye and ear regions of the face, respectively. Whereas the diencephalon is connected to the eye via the optic nerve, the hindbrain (the part of the brain most similar in structure to the spinal cord) forms a larger number of neural connections with all the rest of the face region, especially with the embryonic ear.

The stroma in the hollows between contiguous, expanding segments of the brain is hindered in growth and so becomes compressed and stretched. This stretched tissue forms a system of anchoring bands. These represent the initial, more powerful parts of the meninx and are called *dural bands* or dural girdles (Figs. 6.7–6.9). Toward the crown of the head, the anchoring bands fan out in broad sheets, in conformity with the greater surface growth of the brain in this same region. There is an unpaired and a paired (left and right) dural band. The unpaired band gives rise to the so-called *tentorium* (awning) over the cerebellum whereas the paired dural bands help to form, in part, the *falx cerebri*

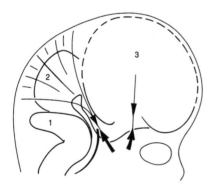

Fig. 6.7. Schematic view of brain of embryo (29 mm long), indicating the forming functions of the brain in the vicinity of the skull. These functions lead to the development of a strong unpaired band of connective tissue between the cerebrum and cerebellum, and a smaller band between the frontal and parietal lobes of each cerebral hemisphere (dural bands). Converging arrows: growth resistance and restraining function of dural bands. 1) cerebellum, 2) midbrain, 3) right cerebral hemisphere.

between the two cerebral hemispheres. It is as if the dural anchoring bands form window-frames around the cerebellum and parts of each cerebral hemisphere. The anchoring band of the cerebral hemisphere divides subsequently into two smaller bands.

Bulging through the windows framed by the stretched dural bands, the brain arches outwards against the skin. Paired brain bulges therefore arise in each of the frontal and temporal regions, and an unpaired bulge in the region of the future cerebellum. The dural bands of each side are interwoven at the base of the brain in the connective tissue

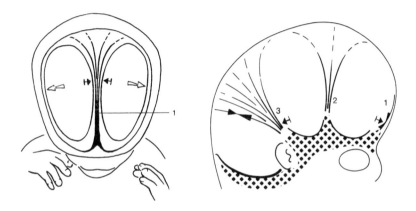

Fig. 6.8. Schematic view of brain of embryo (29 mm long) seen from front, representing left and right dural bands in the frontal region. Converging arrows: growth pressure exerted by the brain on the left and right frontal dural bands. Outlined arrows: growth expansion of both cerebral hemispheres. 1) stretched connective tissue (falx cerebri) from fusion of frontal dural bands.

Fig. 6.9. Schematic lateral view of brain of human embryo (about 29 mm long) showing dural bands (1–3) attached to three elevations in the anlage of the cartilaginous basicranium (densation field, stippled). Converging double arrows: restraining function of stretched connective tissue. Arrows with cross-tails: growth expansion of frontal and occipital lobes of brain. 1) compressed part of frontal dural bands attached to elevation at anlage of crista galli (falx cerebri), 2) dural band attached to elevation of anlage of sphenoid bone, 3) occipital dural band attached to elevation of anlage of temporal bone (tentorium cerebelli).

of the **basicranium** (Fig. 6.10); they therefore represent true girdles (cingula). As the brain grows eccentrically in a dorsolateral direction, the basicranium flattens and this leads to compression of the adjacent stroma on its convex (inferior) side. A basal densation field therefore arises here, external to the dura. These developmental movements are similar to those that we have already encountered in the formation of the axial skeleton. Initially precartilage forms in the densation field and then, in turn, the basicranium becomes cartilaginous. It is only during the fetal period that partially bony structures start to form in the basicranium.

The first bones of the **calvaria** develop at the border of the cartilaginous basicranium. Detraction fields arise here, and in these fields the tissue becomes dehydrated and extremely dense (Chapter 3, Fig. 3.31). Initially, the loss of water from this tissue causes the formation of only microscopically small bony islands (spicules). Whereas growth is at a

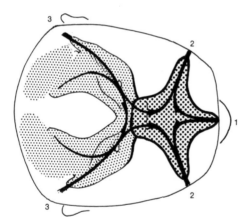

Fig. 6.10. Basicranium of 29 mm long embryo seen from above. Black lines: stretched connective tissue of dural bands interconnecting in basicranium (true dural girdles); adjacent cartilaginous basicranium, stippled. 1) nose and basal attachment at crista galli of dural girdle between left and right cerebral hemispheres (falx cerebri), 2) basal attachment of dural girdle between frontal and temporal lobes at lesser wing of sphenoid, 3) basal attachment of dural girdle between occipital lobes of cerebrum and cerebellum (tentorium cerebelli).

standstill inside a hardened bony island, new bone tissue is laid down radially on its surface. Five foci of ossification arise, one in each of the five large windows in the connective tissue of the skull anlage. These foci enlarge into the five fan-shaped bones of the skull roof (calvaria); they are the anlagen of the left and right *frontal bones* (Fig. 6.11), the left and right *temporal bones*, and the broad unpaired *occipital bone*.

The skull bones move away from each other due to increase in the girth of the brain. This leads to gliding movements in the tissue around the foci of bone formation. Corresponding to the rules of a detraction field, new bone formation continues on the firm substrate of previously formed bone, that is, on the surface of existing skeletal foci (bony spicules and trabeculae). At the tip of a radiating bony spicule, the soft tissue on the surface is dragged away (detracted) like the removal of a tiny

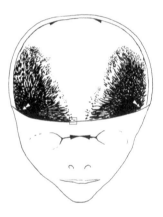

Fig. 6.11. Frontal bones of 50 mm long fetus. The foci of the two frontal bones separate from each other as the brain enlarges eccentrically. The tissue mantle around the bones (periosteum) glides against the bony ground substance while being compressed between the growing brain and the overlying skin. Detraction fields arise here leading to a radial deposition of new bone. Half-headed arrows at top: weak growth resistance of skull (still composed of only soft tissue) against growth pressure from the brain. Converging arrows: restraining function of connective tissue between eyelids (interorbital ligament, see Fig. 5.8). White arrows: original foci of ossification of frontal bones. Box in anteromedial edge of right frontal bone relates to Fig. 6.12.

thimble (Fig. 6.12). By this process, the detraction field increases in size, preferentially at the tips of bony spicules. The radiating construction of the skull bones can be recognized with the naked eye in fetuses that are just three months old. It is only later, when the local directions of the detraction fields become markedly divergent, that the ray-like bony construction is effaced and so-called compact bone forms superficially. The external surface of a skull such as the one depicted in Figure 6.13 consists almost only of compact bone.

Fig. 6.12. Section through a detraction field: a single bony spicule from the anteromedial edge of the right frontal bone (boxed region in Fig. 6.11). The bony matrix formed appositionally in layers, stippled. Bone forming cells (osteoblasts), black. The gliding "bone-skin" (periosteum) is indicated by wavy lines. 1) blunt end of streamlined bony spicule directed (laterally) toward the focus of bone formation, 2) and 3) periosteum. The arrows indicate the opposite directions of gliding of the hard bony spicule and the periosteum.

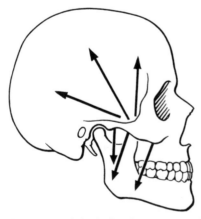

Fig. 6.13. Adult skull indicating muscles of mastication. Arrows designate directions of previous growth dilation of muscle fibers.

Muscles

All muscles differentiate in fields where the cells become extended under specific conditions (dilation fields; Chapter 3). In general, *skeletal musculature* originates in dilation fields associated with the piston-like growth of the cartilaginous skeleton. If one knows the location and the main vectors of growth of the cartilaginous skeletal elements, then one can predict the course and orientation of muscle fibers, muscles, and muscle groups. We will proceed to do this in the following examples.

Muscles of the head region: The location where the muscles of **mastication** arise is determined by the growing skeleton (Fig. 6.13). In turn, the direction of skeletal growth depends on the ascent (**ascensus**) of the brain and the simultaneous descent (**descensus**) of the viscera away from the basicranium. In Chapter 7 we will describe how the diaphragm is flattened by the growth of the liver adjacent to its inferior surface. As this flattening occurs, the diaphragm is increasingly separated from the vertebral column; the diaphragm loses its connection to the thoracic vertebrae but remains joined to the lumbar vertebrae. It is commonly said that "the diaphragm descends." The cervical and thoracic viscera also descend along with the diaphragm. The **hyoid bone**, which is anchored by tension to the lower jaw (*mandible*), is a component of these descending viscera. In the course of the descent, the mandible bends to form an increasingly prominent angle projecting inferiorly (Fig. 6.14). Thus, dilation fields arise where muscles develop in the space between the skull and the mandible. In particular, several muscles of mastication arise in dilation fields between the basilateral wall of the skull and the mandible. Another muscle of mastication (buccinator muscle) differentiates between the cheekbone (zygomatic arch) and the mandible. At the same time, many other muscles are developing in dilation fields in the head–neck region. The *muscles of facial expression*, which arise between skin and skeleton, also develop in dilation fields. In addition to other factors, the location and form of facial muscles depends, on the one hand, on the growth of the skin and, on the other hand, on the piston-like growth of the cartilaginous skeleton (see Chapter 5: Organs in the region of the foregut).

Muscles of the neck region: The distance between the skull and the ventral wall of the thorax increases with the piston-like growth of the cervical vertebrae and the ribs. At the same time, the entire **occiput** (back of head) is arching farther posteriorly. A dilation field arises between the occiput and the sternum, and a prominent oblique muscle of the neck (*sternocleidomastoid muscle*) differentiates in this field (Fig. 6.14). In the embryo, the principal direction of the extension of this muscle is between the anterior tip of the first rib, effectively the sternum, and the dense (**petrous**) part of the temporal bone behind the ear.

As the occiput becomes higher and wider due to the growth of the brain, a group of muscles (*posterior cervical musculature*) develops in a dilation field that arises in the deep cervical region between the occiput and the vertebrae of the neck (Fig. 6.15). Here, the muscle fibers diverge superiorly, corresponding to the vectors of the piston-like growth of the cartilaginous skeleton of the occipital region.

Muscles of the back region: An especially interesting muscle is the so-called *trapezius*. For a long time, this muscle was called the "cowl" (*cucullaris*) because of its similarity to a monk's hood. The trapezius is the most superficial of all the muscles in the region of the head, neck, and back (Fig. 6.16). This muscle arises relatively late in development because the skeletal components that are required for its formation take a long time to mature. The trapezius muscle does not appear until the last condensations of the cervical vertebrae (i.e., their spinous processes) form under the skin of the embryo's back. The trapezius then develops to cover all the muscle layers that formed previously. The trapezius is a beautiful paradigm for the positional development of muscle belly and tendon. The dilation field for the trapezius arises according to both the elongation of the cervical and thoracic vertebral column, and the increase in the transverse distance between the shoulder and the vertebral column. On the other hand, the flat mirror-like tendons (aponeuroses) of the trapezius muscle arise in zones where the skin lies close to the skeleton, namely at the skull, at the spinous processes of the lower cervical and upper thoracic vertebrae, at the 12th thoracic vertebra, and over the spine of the scapula (Fig. 6.17). The different

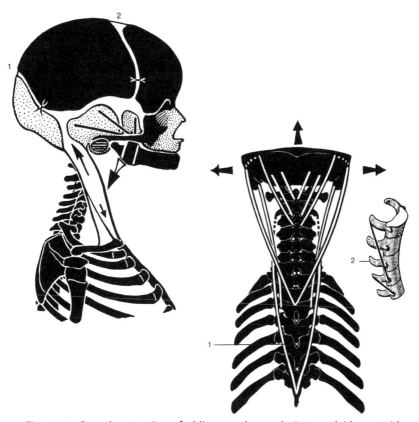

Fig. 6.14. Growth extension of oblique neck muscle (sternocleidomastoid muscle, diverging arrows) and formation of the increasing angle of lower jaw (outlined arrow below mandible) during descent of the viscera. The sternocleidomastoid muscle develops during the growth elongation of the vertebral column in a dilation field between the ventrally growing ribs and the dorsally growing occiput. The muscles of mastication are represented according to their principal fiber directions. 1) occipitoparietal (lambdoid) suture, 2) frontoparietal (coronal) suture.

Fig. 6.15. Dilation fields of deep neck musculature (white lines) between the elongating vertebral column and the transversely enlarging occiput (arrows). 1) cross indicates spinous process of 4th thoracic vertebra arising from union of left and right vertebral (neural) arches. Inset (viewed from anterior right side) shows earlier stage in dorsal growth of a cartilaginous vertebral (neural) arch (2); dilation fields between growing tips of neural arch and transverse process indicated by lines.

directions of the muscle fibers in this extremely broad muscle are associated with markedly different voluntary movements.

In deeper regions of the back, muscles arise in direct relation to the growth of the cartilaginous vertebral column and the ribs. Certain long, cord-like muscle bundles (erector spinae group) originate here, as well as a large number of short, more obliquely oriented muscles (Fig. 6.18).

Muscles of the ventral body wall: The muscles of the ventral body wall also differentiate in dilation fields. One can distinguish an inner layer and an outer layer of **intercostal muscles**. The ribs play the major role in determining the location and direction of these muscles. Dilation fields arise between the growing tip at the ventral end of any one rib and the older, firmer portion of the adjacent (upper and lower) ribs.

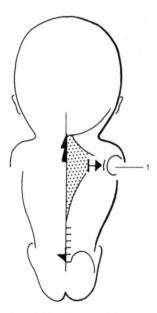

Fig. 6.16. Human fetus 40 mm long. Dilation field of right half of trapezius muscle (most superficial back muscle) stippled. Upper and lower half arrows: growth elongation of cartilaginous vertebral column. Arrow with cross-tail: transverse broadening of the shoulders. 1) shoulder (glenohumeral) joint.

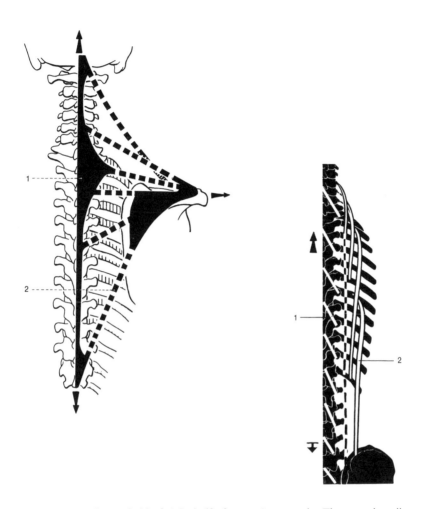

Fig. 6.17. Dilation field of right half of trapezius muscle. The muscle cells are dilated (dashed lines) between four tendinous zones (shown in black, the so-called tendinous "mirrors"). Arrows indicate main growth movements. 1) central tendinous "mirror" over the spines of lower cervical and upper thoracic vertebrae, 2) muscle bundle in dilation field.

Fig. 6.18. Dilation field of deep back musculature (erector spinae muscles). Arrows indicate growth elongation of the cartilaginous vertebral column. 1) short oblique (transversospinal) muscle, 2) long straight muscle (iliocostalis muscle).

The intercostal muscles differentiate in these dilation fields. Between any two adjacent ribs, there must be two dilation fields. Due to the conical shape of the growing thorax, the external dilation field must be inclined obliquely downwards and the inner one must run in the opposite direction, obliquely upwards (Fig. 6.19). Thus, from a phenogenetic viewpoint, we find two sets of intercostal muscles, internal and external. The external musculature is always dilated by the growth of the rib with less curvature, that is, the next lower rib.

The chest musculature continues into the abdominal region as the so-called *abdominal wall musculature*. Once again, the adult muscular patterns overlap and cross in this region according to the positions and directions of their growth dilation fields. These muscle systems connect the thorax with the pelvis and several have tendons that cross the midline of the abdomen. The longest is the so-called straight abdominal muscle (*rectus abdominis muscle*), which owes its length to the tipping movement of the pelvis during elongation and straightening of the embryo; in the course of this movement, an especially long dilation field arises in the ventral wall of the abdomen (Fig. 6.20).

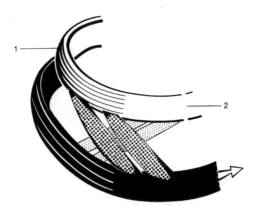

Fig. 6.19. Ascending and descending oblique muscles in dilation fields between adjacent growing ribs (intercostal muscles). 1) longitudinally lined region indicates older portion of upper rib, 2) white zone indicates younger portion of same rib. Outlined arrow indicates direction of piston-like growth of younger portion of adjacent lower rib. External intercostal muscle, coarse stipple; internal intercostal muscle, fine stipple.

The Limbs

Early development: The limbs provide us with an excellent example of the biodynamic factors of phenogenesis. Detailed investigations of human embryos have revealed the following.

The *limbs* arise from the ectodermal ring of the lateral body wall at unique sites where the **peritoneum** of the body sac lifts away from the bed of the spinal cord to establish a small angle of firm inner tissue between peritoneum and spinal cord (Fig. 6.21: 1 and 4). These angles, two on each side, arise because the spinal cord is elongating rapidly and moving dorsally, whereas the peritoneum with the viscera is remaining relatively "short" because it is anchored ventrally to the connecting stalk and yolk sac. As there are four such angles, four limbs will differ-

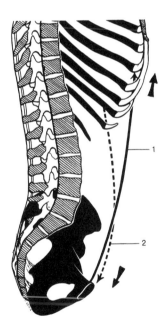

Fig. 6.20. Dilation fields of anterior abdominal wall. Large arrows indicate preceding growth displacement between thorax and pelvis. 1) straight abdominal (rectus abdominis) muscle, 2) oblique descending abdominal muscle (external abdominal oblique muscle).

entiate: two upper and two lower. The significance of these angles was not appreciated until the advent of regional comparison based on total reconstructions of human embryos. The inner tissue of these angles is usually traversed by veins whose stems in the vicinity of the peritoneum remain short but whose tributaries to the spinal cord elongate. Like all

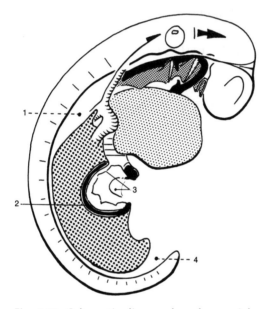

Fig. 6.21. Schematic diagram based on serial section reconstruction of 2.57 mm long human embryo (Carnegie Stage 12). The unequal longitudinal growth of the spinal cord and the peritoneum leads to the formation of an upper (1) and a lower angle (4) between the tissue bed of the spinal cord and the peritoneum on each side. Ectodermal (limb) placodes arise in the ectodermal ring over these four angles; the growing skin piles up at the dorsal edge of each limb placode to form the limb fold. Pericardium light gray stipple, peritoneum coarse stipple, foregut dark-gray stipple, aorta black. 1) upper angle between spinal cord and peritoneum (site of formation of upper limb), 2) body sac (coelom) lined by peritoneum, 3) endoderm cut at junction with yolk sac, 4) lower angle between spinal cord and peritoneum (site of formation of lower limb). Arrow: growth movement of brain. Half arrow: restraining function of head vein (superior cardinal vein). Note the nasal gubernaculum as the superior continuation of the pull of the connective tissue of the dorsal aorta, also illustrated in Fig. 2.36.

embryonic vessels, the stems possess a tensile strength and grow less rapidly than their delicate terminal branches. Wherever the powerful veins exert a restraining action, the skin becomes even thicker than the adjacent parts of the ectodermal ring. This thickened ectoderm constitutes the so-called *limb placode*. The placode becomes densely innervated (see Chapter 4). The embryonic skin dorsal to the placode rises away in a distinct arch.

The first elevations that we call limb folds are therefore really skin folds. These folds arise because the ectoderm is gliding off the back of the embryo in the direction of the abdomen. The surface area of each limb fold increases along with the surface growth of the entire ectoderm. The ectoderm is therefore the motor driving limb development. During this growth, the limb fold flattens and inclines anteriorly (Fig. 6.22). If one seeks to determine why the upper limb comes to bend forwards and also, by further growth, comes to lie in the vicinity of the mouth, then one can establish the following facts. The ectoderm of the limb fold braces itself on the resistance of the adjacent trunk wall as

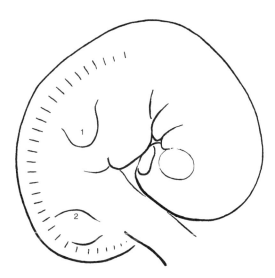

Fig. 6.22. Drawing from total reconstruction of 6.3 mm long human embryo (Carnegie Stage 14). Anlagen of the upper limb (1) and lower limb (2) are distinctly different.

it grows continually in surface area. The ectoderm of the trunk region represents a kind of firm "sleeve hole" to the rim of which the limb fold is "stitched." The rim of this "sleeve hole" sinks into the body wall opposite the site where the wall of the body sac with its peritoneum provides a less firm substrate. As this weaker region is the ventral portion of the "sleeve hole" (the anlage of the armpit or axilla), the entire limb fold inclines anteriorly over this part of the body wall. Thus the low relief of the axilla and the high relief of the limb fold act like a torque-couple. The growth tilting represents a relative approximation of the limb anlage toward the ventral body wall and is therefore a growth adduction. This adduction occurs well before there is any kind of differentiation of skeletal or muscular tissues in the limb.

At this stage of limb fold development, it is already possible to distinguish a flexor surface closely apposed to the chest wall, and an outer, extensor surface. The thickened ectoderm on the flexor side of the limb fold becomes so strongly innervated that, by the beginning of the 2nd month, most of the inner tissue of the limb consists of nerves. Vascular branches from the large blood vessels near the heart are also sprouting into the adjacent limb folds. The stems of these vessels remain short and so rein-in the growing limbs. This restraining action of the early blood vessels leads to a more pronounced asymmetry in the growth of the young limb folds, and bends (flexions) start to arise along the limb fold. Thus, the limbs are already segmenting before the skeleton and the musculature have developed. The early segmentation is tripartite, consisting of the arm, forearm, and hand in the upper limb, and the thigh, leg, and foot in the lower limb. Gradually the bends between the segments become more distinct. On the flexor surface, the thickening of the skin becomes more noticeable, whereas on the extensor surface, the skin remains thin.

Already by the start of the 2nd month of development, on account of their asymmetrical growth, the tiny arm and leg anlagen are expressing growth functions, in the sense of incipient grasping and stepping movements.

Even in adults, the calloused skin on the palm of the hand and on the sole of the foot is still remarkable. As elsewhere, the skin here has also acquired a dense innervation and a rich blood supply on account of the thickening of the embryonic ectoderm. The cutaneous flushing seen on the palms and soles of newborn children is just as much a consequence of early development as is the unique sensitivity of these regions. It was once believed that human limbs passed through a "fin stage"; however, such a stage cannot be demonstrated in human ontogeny. The so-called paddle stage of hand development is not a sign of a recapitulation, but is simply a characteristic feature of the early surface growth of human limb ectoderm.

The formation of the limb skeleton: Investigations on the developmental movements of the upper limb show that initially it grows laterally, forming most of the future arm. As the limb elongates, it makes a forearm, which bends against the arm. Then the forearm elongates, and in so doing makes the hand, which bends against the forearm. The above growth movements are associated with the development of a succession of undulations along the free margin of the limb fold (Figs. 6.23, 6.24). The first wave-like elevation initiates the formation of the

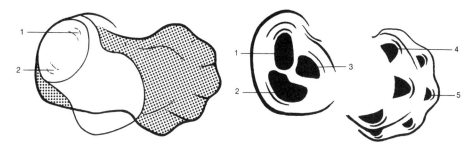

Fig. 6.23. Superimposed stages in development of anlage of right upper limb (8.1 mm, 10 mm, 15.5 mm long embryos). Development of a grasping movement during embryonic growth adduction of upper limb and growth extension of hand. 1) site of development of radius in youngest embryo (8.1 mm), 2) site of development of ulna in same embryo.

Fig. 6.24. Annotation to Fig. 6.23, showing how the skin initiates the formation of the skeletal densation fields under different sequential skin crests for: 1) humerus, 2) ulna, 3) radius, 4) row of three wrist (carpal) bones, 5) row of four wrist (carpal) bones. The densation fields for bones of the midhand (metacarpus) and fingers arise later.

densation field for the humerus deep in the inner tissue of the arm. The 2nd and 3rd elevations initiate the formation of the two bones of the forearm (ulna and radius). Additional crests lead to the formation of densation fields and then cartilages of the wrist (carpus) and midhand (metacarpus). Along the free margin of the midhand, the skin undulates in a series of five crests with intervening troughs. On the crests, the terminal ectoderm of the limb is thicker than in the vicinity of the troughs. Initially, the crests are elevations of skin possessing divergent wedge-shaped epithelial cells. The crests develop gradually into digits (Fig. 6.25) according to the properties of a diverging wedge epithelia (Chapter 3, Fig. 3.7). Skeletonization occurs subsequently in the inner tissue of the digits.

At all stages, the ectoderm on the extensor side of the limb anlage is thin and the adjacent inner tissue is rich in fluid. Conversely, on the flexor side, we find that the ectoderm is thick and the inner tissue is highly cellular; these inner tissue cells are aligned perpendicularly to the ectoderm. Seen from a physical perspective, this means that in both ectoderm and inner tissue, the cells are exerting a reciprocal growth pressure on one another. The first vascular networks of the anlage of the developing limb arise along the undersurface of the ectoderm, particularly on the flexor side. These vessels carry not only raw building materials to the ectoderm, but also simultaneously convey metabolic

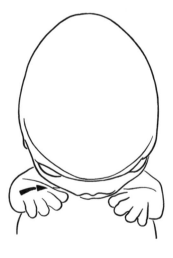

Fig. 6.25. Human embryo 20 mm long. Inward rotation (adduction, flexion, and pronation) of the upper limb during growth (arrow) in the sense of an incipient grasping movement.

by-products, especially water, away from the inner tissue of the limb anlage. With increasing dehydration, a focus of cellular consolidation arises in the interior of the limb stroma at a definite distance from the overlying skin; this focus is the anlage of precartilage. Cartilage arises within this focus and, later, bone. The rules governing these differentiations were described in Chapter 3. We can summarize the above by stating that the skin, particularly its ectodermal component, is the motor or main formative apparatus for the shaping of a developing limb. The young skin provides the blueprint, as it were, for the skeleton.

When the blood vessels growing into the limbs have become sufficiently numerous and strong, they act, as do all vascular stems and larger conducting channels, as restraining structures with shaping functions (Figs. 6.26, 6.27). As they become tenser, the arterial stems in the subcutis (hypodermis) acquire another special capacity, namely

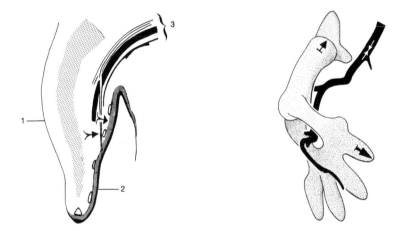

Fig. 6.26. Upper limb of 8 mm long embryo in longitudinal section. Restraining function of neurovascular bundle (converging half arrows at 3) leads to closer approximation of hand and trunk (growth adduction). Early growth bending initiates the formation of joints. 1) thin skin on extensor side of upper limb, 2) thick skin on flexor side of upper limb (and in region of axilla). Tailed arrows: metabolic movements. Anlage of skeleton, light stipple.

Fig. 6.27. Human embryo 10 mm long. Inward rotation of forearm (flexion and pronation) due to restraining function of artery of upper limb (converging double arrow) and piston-like growth of cartilages (diverging arrows).

the ability to express increasing amounts of fluid into their connective tissue beds. The rounded profile of the skin in the suckling infant, which is so well portrayed in the chubby putti of baroque sculpture, is based on the tumescence of the hypodermis that has developed in the above manner. The "reining-in" action of the large neural and vascular structures on the flexor side of the limb initially forces the limb to grow with a curvature (flexion). Then a more distinct kinking arises and following this, the first segmentations of the young limb. Once again, the example of limb development illustrates how all differentiations are correlated dynamically through growth. While the major arm vessel (brachial artery) is remaining short on the flexor side of the upper limb, the forearm is becoming flexed at the elbow and inwardly rotated (pronated), in such a way that the hand grows into a gripping position in the vicinity of the mouth (Fig. 6.25). If the embryo lacked the capacity to develop its hand in a gripping position, then the prerequisites would not be available for the child to learn to grip voluntarily.

The formation of limb muscles: The following generalization is also valid for the limbs: If one knows the development of the form of the skin and the vectors of piston-like growth of the cartilaginous skeletal segments, then one can deduce the locations and directions for the dilation fields of adjacent muscles. Genetic information alone is quite insufficient to explain muscle phenogenesis (i.e., the formation of actual muscles). In order to enable such differentiations, phenogenetic processes, with rules of their own, must interact with the genetic material. This phenogenesis is the basis for the particular anatomy found in adults.

Some examples of this unique phenogenesis will now be given. The growth of the upper limb is accompanied by the development of joints, as well as muscles whose tendons cross over the joints. The particular arrangement of muscles and tendons is prepared for by growth processes. The arm musculature originates in the elongation of the cartilaginous skeleton of the arm and the growth of its joints (Fig. 6.28). The forearm musculature develops in a similar way with the growth of its cartilaginous skeleton and its joints (Figs. 6.29, 6.30). Arm and forearm muscles differentiate in dilation fields, which have various widths and directions

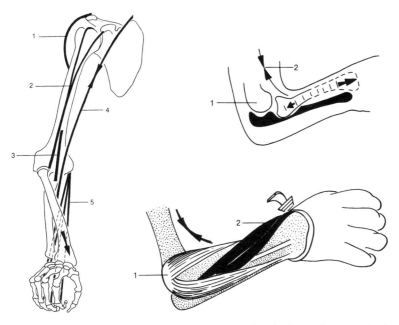

Fig. 6.28. Directions of important dilation fields of adult right upper limb. Arrow designates preceding growth elongation of radius. Converging double arrow: restraining function of brachial artery (4). 1) lateral shoulder (deltoid) muscle formed by growth adduction of arm; 2, 3, and 5) various flexor muscles that arise during growth elongation of skeleton (2) biceps brachii muscle, 3) brachialis muscle, 5) flexor digitorum superficialis muscle).

Fig. 6.29. Inward rotation of right forearm (lateral view) showing the basic relations between the three skeletal segments that constitute the elbow joint during growth flexion and pronation. Diverging double arrow: growth elongation of radius. Converging double arrow: brachial artery (2) remaining relatively short. 1) humerus. Ulna black.

Fig. 6.30. Characteristic dilation fields in growing right forearm of fetus at the start of the 3rd month (about 30 mm long). The fields are interpreted according to the piston-like growth of the cartilaginous skeletal segments (stippled) in relation to blood vessels. Black double arrow: restraining function of brachial artery. Outlined arrow: inward rotation of the radius. 1) attachment site on humerus (lateral epicondyle) for posterior forearm muscles (extensor carpi radialis longus and extensor carpi ulnaris muscles), 2) oblique ("button-hole") muscles of thumb (2 indicates abductor pollicis longus muscle; extensor pollicis brevis muscle lies below).

at different sites between the skeleton and the skin. Thus there exist arm muscles of varying lengths, widths, and orientations. In the course of growth, muscle phenogenesis is an important prerequisite for the subsequent use of muscles, under the influence of the brain, for conscious movements and voluntary actions. What is absolutely remarkable is that the body has only a finite number of muscles at its disposal to accomplish its immense variety of arbitrary movements, gymnastic skills, and manual dexterity. The initial step in comprehending human movement is to appreciate that these capacities are founded on the complex growth movements of the embryo's limbs.

In the lower limb, we find developmental movements analogous to those in the upper limb. Once again, a large blood vessel acts as a restraining apparatus. And again, there is the piston-like growth of the young skeleton, but here acting under different initial conditions. In the lower limb, the major blood vessel has an alignment corresponding to the number 1 in Figures 6.31 and 6.32. As is indicated by the outlined arrow in Figure 6.31, an inward rotation of the growing thighbone (femur) occurs about the axis of the major vessel. This leads to the formation of many quite different dilation fields, above all, to the musculature of the *extensors, flexors, adductors,* and *abductors* of the thigh.

The muscles that are known as adductors (dashes, Fig. 6.31) are attached posteriorly—according to their growth dilation—to the skeleton of the femur. The inward growth rotation of the skeleton here implies an oblique dilation field posterior to the artery of the thigh (femoral artery). Accompanying the growth adduction of the thigh, a dilation field develops for the powerful *hip musculature* (gluteal muscles) on the external aspect of the hip. All the various dilation fields can be demonstrated long before the child learns to stand.

The restraining function of the embryonic femoral artery is coupled with a growth bending (flexion) of the lower limb at the knee joint; the leg is now distinct from the thigh. Once again, this bending is combined with the development of both extensor and flexor groups of muscles (Fig. 6.32). The extensors arise hugging the femur, while the flexors are displaced relatively far from the skeleton. No two muscles are alike. All fully developed muscles are different because they arise

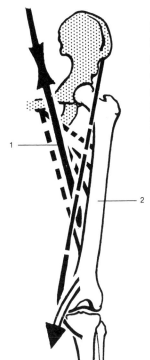

Fig. 6.31. Principal directions of major dilation fields in adult thigh (left lower limb, anterior view); pelvis stippled. Converging double arrow: growth resistance of femoral artery (1). Lower arrow: growth elongation and adduction of femur (2) oblique to femoral artery. Dilation fields of inner (adductor) muscles of thigh indicated by short dashes. Dilation field of sartorius muscle indicated by long dashes.

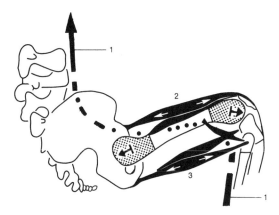

Fig. 6.32. Dilation fields in thigh region of 32 mm long fetus (right thigh, lateral view). Still-cartilaginous parts of thigh skeleton stippled. Diverging arrows with cross-tails: piston-like growth of cartilage. Musculature black. White arrows: growth dilation of muscles. 1) growth pull due to sequence of large arteries (between abdominal aorta and posterior tibial artery indicated by dot–dash line), 2) extensor musculature (quadriceps femoris muscle), 3) flexor musculature (a "hamstring" or long head of biceps femoris muscle).

under unique initial conditions at each site (Fig. 6.33). More or less powerful muscles develop in dilation fields according to the greater or lesser distance, respectively, that the muscle can occupy between the skeleton and the skin at any particular site.

All the above growth dilations are coupled with slight but characteristic gliding movements. These arise because the direction of growth extension of the muscle fibers is invariably inclined to the main direction of the **tensile** (extending) **stress** (Fig. 3.28). The gliding movement

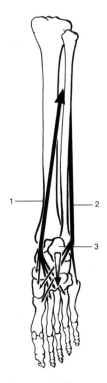

Fig. 6.33. Main directions of dilation fields of deep calf musculature in adult right leg (viewed from below and behind). Outlined arrow: growth elongation of the former cartilaginous skeleton. 1) so-called elevator muscle of medial margin of foot (tibialis posterior muscle producing inversion), 2) so-called elevator muscles of outer margin of foot (peroneus longus and peroneus brevis muscles producing eversion), 3) heelbone (calcaneus).

of muscle fibers is of significance for the "porosity" of the muscle, in that it provides space for a substantial **vascularization**. Tendons arise where such spaces fail to develop and a transverse compression exists instead. Thus many tendons in the body are found in regions where the skin lies close to the skeleton. If these tendons are subject to a particularly strong transverse pressure at a given site, then a contusion field may arise within the tendon as happens in the formation of the young knee-cap (patella) and other **sesamoid** bones.

Joints

The following factors contribute phenogenetically to the formation of human joints. The slow growth (growth extension) of the inner tissue on the extensor and flexor sides of the limb anlage, in opposition to the piston-like growth of the cartilage, leads to the formation of slings of stretched inner tissue along the major traction lines in the tissue bed of the skeleton (Fig. 6.34). These are the so-called **articulation slings**. These tissue tracts, which are dilated proximally on the extensor and flexor sides by the piston-like growth of the cartilage and distancing of the ectoderm, merge together in the distal regions of the limb, so forming continuous loops. Where the distance between the skeletal anlage and the skin is large, as occurs on the flexor side of the upper limb, the dilated segment of the sling is thicker than on the extensor side (Figs. 6.34, 6.35). Thus the flexor muscles developing in this dilated segment are more powerful than the extensor muscles in the other dilated segment. The formation of these slings in a limb is a necessary prerequisite for flexors and extensors interacting as so-called *antagonist*

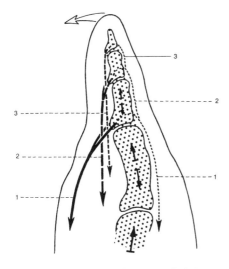

Fig. 6.34. Schematic diagram of skeleton of hand and 5th digit in human embryo (about 15.5 mm long) showing different joint slings (articulation slings). The flexors (solid black and dashed lines) and extensors (dotted line) constitute the arms of these slings, paired respectively as 1 & 1, 2 & 2, 3 & 3. A joint socket arises under the apex of each of these three slings. Diverging double arrows with cross-tails: piston-like growth of cartilage. Outlined arrow: start of growth flexion of finger in correlation with stronger growth pull of more massive flexor muscles (on flexor side: 1) flexor digiti minimi muscle, 2) flexor digitorum superficialis muscle, 3) flexor digitorum profundus muscle).

muscles (opposing partners). If the thick musculature on the flexor side is stressed on stretching by the piston-like growth of the eccentrically located skeleton, then the weaker anlage of the extensor musculature yields more easily. The greater thickness of the flexor muscle anlage, as opposed to the extensor, signifies that the flexor offers a greater growth resistance to the piston-like growth of the skeleton. This asymmetry initiates the formation of joints. The flexing of the skeleton increases at sites beneath (i.e., proximal to) the apex of each articulation sling. A gradual shearing occurs, leading to discontinuity in the skeletal anlage and eventually to the formation of a *joint space* containing a fluid similar to egg white (**synovial fluid**; Figs. 6.36, 6.37).

We can therefore now understand why muscles always cross over one or more joints. Seen from the perspective of developmental dynamics,

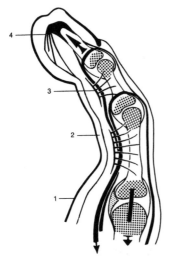

Fig. 6.35. Finger of 59 mm long fetus showing retaining bands (annular parts of tendon sheath) around powerful flexor tendons. These bands arise through the stretching of connective tissue as the flexor tendons attempt to lift away from the skeleton during growth flexion of the finger. Cartilaginous parts of skeleton stippled. Upper arrow: growth elongation in distal phalanx. Lower arrow with cross-tail: growth pressure against metacarpal bone. 1) epidermis, 2) dermis (corium), 3) retension band developing in an arc around flexor tendon (pars annularis of flexor tendon sheath), 4) detraction field at end of finger (terminal phalanx; refer to Fig. 3.30).

muscles cross joints because the muscles develop in segments of various large sling systems and joint spaces arise within the compass of these same slings.

Fluid is expressed into the joint space, just as occurs in other spaces formed by tissue gliding (e.g., inside the sheaths of tendons). With the accumulation of synovial fluid in the joint space, the tissue at the periphery of the space becomes stretched forming the *joint capsule*. Those parts of the joint capsule that are particularly well stretched are called *ligaments* (Fig. 6.37). Fundamentally, there are no differences between the organs of movement in the head, the trunk, and the limbs.

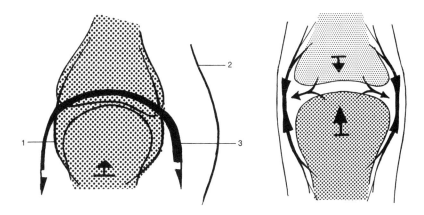

Fig. 6.36. Formation of joint under apex of articulation sling. The more powerful growth pull of the flexor (on right) causes a shearing of the cartilaginous tissue under the apex of the articulation sling thereby forming a joint socket sliding over a joint ball. 1) perichondrium (anlage of joint capsule), 2) epidermis on flexor side of joint, 3) flexor crossing over via articulation sling into extensor (on left). Half arrows: growth pull. Arrow with cross-tail: piston-like growth of cartilage.

Fig. 6.37. Schematic diagram of a joint space with accumulation of fluid (tailed arrows). The former perichondrium (labeled 1 in Fig. 6.36) is stressed by both transverse pressure and longitudinal pull so that it develops into a tense joint capsule (converging double arrows). Arrows with cross-tails: piston-like growth of cartilage.

Chapter 7

Viscera

The **viscera**, in distinction to the outer organs of the body that are
mainly derived from the skin, are those organs whose functions
are initially only accessible to investigation underneath the skin. It is
customary to distinguish viscera of the head region (cranial viscera),
viscera of the neck region (cervical viscera), and, in adults, the especially large internal organs of the thorax and abdomen (thoracic and
abdominal viscera; Fig. 7.1). The superior and inferior parts of the abdomen are designated the *abdomen proper* and the *pelvis*, respectively, and
the organs of the posterior wall of the abdomen behind the peritoneum
are designated **retroperitoneal organs** (Fig. 7.2). There is some basis for
describing the viscera in this order, since visceral development in the
superior part of the body occurs at an earlier stage than in the inferior
part.

The fact that an artist can characterize a whole personality in a portrait may have its basis in ontogeny. In early development, almost the
entire human body is manifested by the anlage of the head. This does
not imply that ontogeny repeats a phylogeny in which, say, our prede-
cessors consisted only of heads. Rather, it tells us that head development takes precedence in the development of the human embryo. It
would be naïve to claim that the brain must represent, on the basis of
its early ontogeny in humans, a phylogenetically ancient organ. The
excavation of fossil species that lived millions of years before us has
shown that no stages ever existed that anticipated human differentiation. The investigation of human ontogeny permits conclusions about
only those factors that lead to the differentiation of an adult human

organism. However, in themselves, these factors cannot shed any light on the preceding phylogenetic history of the human race!

Ontogenetically, we can determine the following. The organs of a human embryo 10–20 mm long are already so well developed that we can designate them using the same anatomical nomenclature that is applied to adults. Of course, the proportions of the embryo's body are quite different compared to the adult (Figs. 7.1–7.4). The trunk of the embryo is scarcely wider than the head, which the embryo holds well

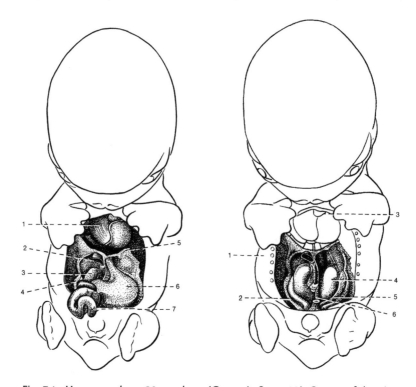

Fig. 7.1. Human embryo 20 mm long (Carnegie Stage 20). Survey of the viscera. 1) heart, 2) inferior vena cava, 3) right suprarenal gland, 4) mesonephric (Wolffian) duct = outflow channel of embryonic kidney (mesonephros), 5) diaphragm, 6) stomach, 7) intestinal loops.

Fig. 7.2. The same 20 mm embryo as in Fig. 7.1. 1) hypodermis, 2) mesonephric (Wolffian) duct, 3) left atrium (almost completely hidden), 4) kidney, 5) genital gland (gonad), 6) mesentery.

flexed over the heart bulge. The small hands are applied to the chest, which is mostly occupied by the heart. The heart beats so strongly that the body wall moves rhythmically in synchrony.

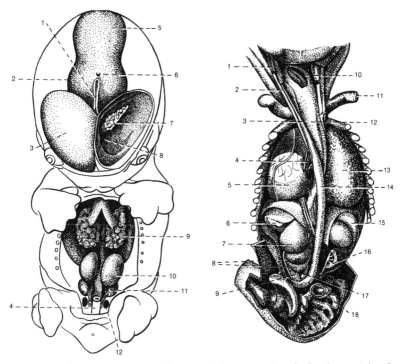

Fig. 7.3. The same 20 mm embryo as in Fig. 7.1. 1) roof of 3rd ventricle of brain, 2) interbrain (diencephalon), 3) right cerebral hemisphere, 4) hindgut (rectum), 5) midbrain (mesencephalon), 6) pineal gland, 7) inward-buckling medial wall of forebrain (choroid plexus), 8) opening between left and right forebrain ventricles (interventricular foramen), 9) lung, 10) suprarenal gland, 11) kidney, 12) left umbilical artery.

Fig. 7.4. Viscera of 24 mm long human embryo (Carnegie Stage 22) seen from behind. 1) cranial nerve XI (accessory nerve), 2) cranial nerve X (vagus nerve), 3) common carotid artery, 4) left pulmonary artery with bronchi (lung removed), 5) heart, 6) spleen, 7) suprarenal gland, 8) kidney (still lobulated), 9) embryonic kidney (mesonephros), 10) opening cut in roof of pharynx revealing entrance to larynx, 11) collar-bone (clavicle), 12) gullet (esophagus), 13) right lung, 14) aorta in thorax, 15) right lobe of liver, 16) right ureter with adjacent renal calices, 17) left ureter, 18) intervertebral disc.

In the chest, the lungs are relatively small; in the abdomen, the liver is the largest organ. The coils of intestine lying under the liver do not yet occupy much space. The retroperitoneal organs located at the posterior abdominal wall are relatively large in comparison with the anlage of the intestine. Among these, the **suprarenal** (adrenal) **glands**, which are larger than the kidneys, are especially conspicuous in early stages. The genital glands (gonads) lie on the lateral borders of the suprarenal glands and extend from the diaphragm (which is still located in the neck region) into the pelvis. The urinary bladder and the rectum fill almost the entire pelvis.

The brain is the most massive and, in its development, the most advanced organ of the nervous system at this stage. Similarly, the heart is a powerful growth center for the blood circulation, and the liver is the largest mass of all the abdominal viscera. Insofar that the liver is an organ of vascular tributaries feeding the heart, the liver is directly subservient to the heart. The heart in turn is responsible for supplying blood to the brain and so is subordinate to the brain. It is therefore apparent that the capacity both to regulate and to be regulated is a fundamental phenomenon of development.

Heart Development

The development of the heart is a sign of early embryonic activity. The heart is more than just a pump; no one would conceive of a pumping organ being built like a heart. The manner in which the adult heart is constructed cannot be understood from a technical viewpoint alone. For this reason there have been repeated attempts to interpret the construction of the human heart from an historical viewpoint. To this end, hearts from different species were ordered according to their similarities, although it was of course appreciated that such similarities are not based on any direct relationship of these organs to one another. Certainly the human heart is similar to the hearts of other animals, but the similarities themselves are not based on a direct connection between the organs. The more significant question is: How does a heart actually arise during the development of any particular animal and according to what rules does such a heart develop? The declaration of all the similarities on earth sheds not the least light on this problem.

A human *heart* is already differentiating in the superior (cranial) margin of the umbilicus in young embryos not quite 2 mm long at about the start of the 4th week. The heart arises in the posterior wall of the body sac (intra-embryonic coelom). The heart is initially just a hollow fold containing much fluid; only later will the heart conduct blood. Metabolic movements occur in the heart from the anlage of the liver toward the brain. The brain at this early stage is the major consumer (sink) of nutrients in the whole embryo. The heart-fold is broad inferiorly where it lies transversely on the margin of the umbilicus, resting as it were upon this margin known as the **transverse septum**. The heart-fold continues superiorly into the region at each side of the embryonic head (Figs. 7.5, 7.6, 7.7). When viewed from the outside, the fold-like anlage of the heart has the form of a short but wide "**X**." The two lower limbs of the "**X**" are separated from each other by a wider distance than the upper limbs. The lower limbs represent the two *inflow paths* and the

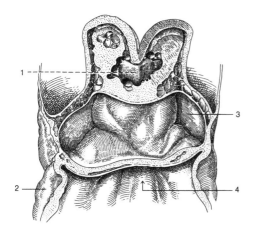

Fig. 7.5. The 0.2 mm long heart of human embryo (about 2 mm long, Carnegie Stage 10) at the beginning of the 4th week of development with X-shaped heart anlage (ventral view). 1) endoderm of mouth membrane (as yet uncorroded), 2) inflow path to heart (right vitelline vein), 3) posterior wall of body sac, 4) entrance to superior intestinal portal. (Davis Embryo, Carnegie Collection, Washington, DC).

upper limbs, the two *outflow paths*. The short *conducting portion* in the middle of the "**X**" between inlet and outlet starts to elongate toward the end of the 3rd week and, by the start of the 4th week, takes the form of a siphon as the so-called *heart loop* (S-shaped heart anlage; Fig. 7.7). The apex of the heart loop initially projects forwards, then more to the right, and later inferiorly. Relative to the conducting portion, the inflow pathway has a more inferior and dorsal location and the outflow pathway is more superior and ventral. The deeper portion of the heart loop is stippled in Figures 7.8 and 7.9, whereas the more superficial portion is shown in solid black. Only the conducting portion constitutes the actual diagonally oriented heart (Fig. 7.7). The conducting portion

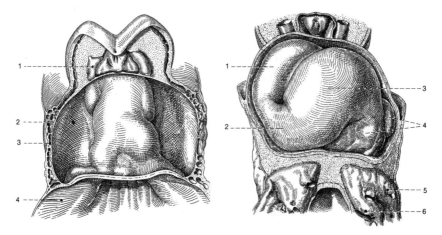

Fig. 7.6. The 0.5 mm long heart of a 2.2 mm long human embryo (Carnegie Stage 10). The heart anlage has become slender and possesses a pair of inflow paths (bottom), a conducting portion (center), and a pair of outflow paths (top). 1) artery of first embryonic flexion fold (1st visceral arch artery), 2) posterior wall of the body sac, 3) mesoderm and mesoblast (loose inner tissue) between amnion and yolk sac, 4) yolk sac wall. (Payne Embryo, Carnegie Collection).

Fig. 7.7. Siphon-shaped heart of a 2.5 mm long human embryo in the 4th week of development (Carnegie Stage 11). 1) outflow path of the heart, 2) terminal part of conducting portion, 3) start of conducting portion, 4) left atrium with afferent vein, 5) left umbilical vein, 6) left vitelline vein (Carnegie Collection).

develops two small cul-de-sacs (diverticula), one on the right and one on the left: these are the anlagen of the right and left **ventricles** (Fig. 7.9). The column of blood within the heart is initially thread-like and lies in an almost vertical direction (Fig. 7.8); the flow seems to shuttle to and fro—"two steps forward, one back." It is only when the inflow volume increases sufficiently and the entire heart loop becomes a regularly pulsating diverticulum in the 4th week that a more unidirectional blood current is established; blood now moves toward the apex of the heart and from there back into the outflow pathway.

Instead of a pump, it is better to envisage the heart as a commutator or switch that reverses the direction and momentum of the blood current between inflow and outflow (Fig. 7.10). Using the right hand, it is possible to mimic the vascular switching in the heart with the movements indicated in Figure 7.11. The so-called right current path (venae cavae → right atrium → right ventricle → lungs) is indicated by the continuous blunter V-shaped line, and the so-called left current path

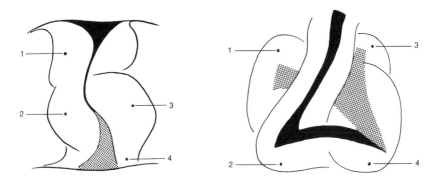

Fig. 7.8. Schematic ventral view of elongated heart showing blood current; the more dorsally located portion is stippled gray and the more ventrally located portion is black. 1) outflow, 2) end of conducting portion, 3) start of conducting portion, 4) inflow path.

Fig. 7.9. Schematic view of "square" or crop-shaped heart. Blood current located dorsally stippled gray, ventral blood current black. 1) right atrium, 2) right ventricle, 3) left atrium, 4) left ventricle. No septa have formed at this stage and the atria and ventricles are still in communication.

(pulmonary veins → left atrium → left ventricle → aorta) by the sharper **V**-shaped dashed line (which should be thought of as lying below the plane of the diagram). We simulate the right current of blood with the thumb, and similarly, the left current with the remaining group of four fingers. Now to mimic the reversal of flow as seen from the right side of the body, we have only to draw our right hand forward, stop it, and then thrust it back with an outward rotation (supination) in a screwing action, so-to-speak with the "flick of a wrist." Doing this quickly makes the reverse switching of the blood current by the heart almost palpable.

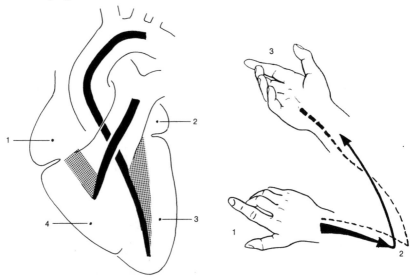

Fig. 7.10. Schematic view of the commutation (reverse switching) of blood flow in the adult heart. Inflow paths stippled gray; outflow paths black. 1) right atrium, 2) left atrium, 3) left ventricle, 4) right ventricle.

Fig. 7.11. Hand movement to demonstrate commutation (reverse switching) by the heart (the so-called "wrist-flick" or "Ruck-Zuck-Schema" of heart action). The movement is viewed from right side and the vertebral column is envisaged to the left of the drawing. The symbol 1) indicates inflow paths with thumb and the group of four fingers signifying the inflow path to right and left side of the heart, respectively; 2) indicates the apex of the heart and the start of rerouting of current paths; and 3) indicates reversed outflow paths (see text).

Investigations in growing embryos have shown blood pressure rises with increasing blood volume. The walls of the heart become stretched with rising blood pressure. The cardiac musculature, which is understandably very powerful in the region of the ventricles, arises in these dilation fields. As blood is propelled into the ventricles eddy-like currents impinge on the cardiac walls, which at this stage consist mostly of *cardiac jelly*. The centrifugal energy of these eddy currents carves out niche-like recesses in the ventricular walls. The niches partially communicate with one another, so that a reticulated relief arises on the inner walls of the heart; the **trabeculae carneae** arise in this relief. The apparatus of the *heart valves* also develops along with this relief (Fig. 7.12). The valve flaps (cusps) differentiate at break-points in the blood flow, that is, at sites of constriction in the siphon-shaped heart tube.

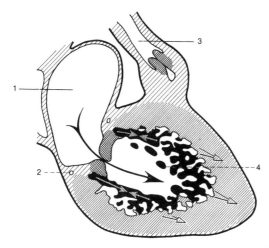

Fig. 7.12. Part of heart lying predominantly on the right in 14 mm long human embryo, viewed from right with vertebral column envisaged to the left of drawing. Tailed arrow indicates main direction of blood flow into the right ventricle. Outlined arrows indicate main direction of growth dilation of right ventricular wall. Cross-hatched arrows indicate retension fields where "heart-strings" (chordae tendineae) arise. 1) right atrium, 2) coronary artery in cross-section, 3) root of pulmonary trunk with semilunar valves, 4) meshwork of musculature (trabeculae carneae) with blood washing through niches below.

The flaps actually function as valves. Abnormal valve structures have been associated with certain types of abnormal blood flow.

Descent of the Diaphragm

We now appreciate that the ascent of the brain and spinal cord (*ascensus*) and the descent of the viscera (*descensus*) are equally fundamental movements in the embryo. Both the ascent with respect to the inferior end of the body and the descent with respect to the superior end are very productive and well-defined movements. If one selects the anterior end of the notochord (a point located in the base of the cartilaginous skull) as a reference point, then the movements of the ascending and descending organs can be determined precisely. The tip of the notochord is the natural quiescent point in the growing embryo and is therefore a fiducial structure to "pin down" the living growth movements.

The descent of the viscera is closely coupled to the development of the diaphragm. The diaphragm is indicated by a thick black line in the median section of a 4.2 mm long human embryo illustrated in Figure 7.13. The diaphragm is attached to the liver (stippled region) posterior

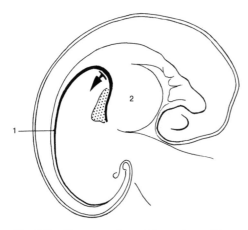

Fig. 7.13. Human embryo 4.2 mm long (Carnegie Stage 13). Start of descent of diaphragm (arrow). Liver (stippled) still high in neck region. Diaphragm black. 1) inferiorly attached part of diaphragm, 2) heart bulge.

to the heart and arches high into the thorax. The inferior end of the diaphragm extends almost to the inferior end of the vertebral column. The segment of the diaphragm between the growing heart and the self-enlarging liver becomes compressed and taut, so that here it will be thin and tendon-like (central tendon of the diaphragm). The diaphragm has always been attached to the liver and as the latter enlarges, the diaphragm flattens against its superior surface. The descending segment of the diaphragm "peels away" increasingly to the front and thus moves farther from the vertebral column (Figs. 7.14, 7.15). It is only in the lumbar region that the diaphragm remains firmly anchored to the

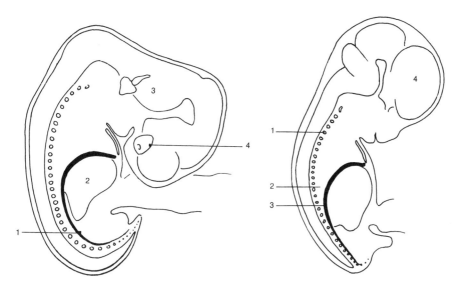

Fig. 7.14. Human embryo 10 mm long (Carnegie Stage 17). The diaphragm (black line) has descended relative to the cranial end of vertebral column (open circles). 1) diaphragm in lumbar region, 2) liver, 3) hindbrain with otic vesicle, 4) eye.

Fig. 7.15. Human embryo 29 mm long (Carnegie Stage 23). The embryo has become relatively narrow in the neck region during the extension of its trunk and the descent of its heart–liver mass. 1) vertebral column, 2) angle between vertebral column and diaphragm (vertebrodiaphragmatic recess), 3) inferior part of diaphragm, 4) right cerebral hemisphere. Compare to Fig. 7.14.

vertebral column. As the liver flattens and the diaphragm descends, the thorax widens in association with the growth of the ribs. This whole process is important for the development of the lungs.

The Formation of the Major Glandular Organs: Liver, Lungs, Thyroid

Liver: While the heart is developing in the 1st month as the main vascular organ, the endodermal cells of the superior umbilical margin are multiplying close to the heart at the site of confluence of the large nutrient-bearing veins. The margin of the umbilicus is also the region where the yolk sac changes into the intestinal wall. The locally increased growth of the endoderm in the superior margin of the umbilicus (transverse septum), which is a particularly vascular region, leads to the formation of the *liver*. The liver also receives a contribution from the epithelium of the **serosa** lining the adjacent body sac. The liver therefore grows directly under the heart in the superior umbilical margin in a powerful suction field lying between the skin, the inflow vessels to the heart, and the serosa (Figs. 7.16–7.18). This suction field is associated with the pulse-beat as the heart drives blood into the aorta. The apex of the heart lifts with each beat, enhancing the suction field in the region behind the heart. The liver develops in this particularly strong suction field as a large glandular organ, in conformity with the general rules for suction fields described in Chapter 3.

As can be shown by a comparison of different developmental stages, the endoderm of the intestinal wall at the superior margin of the umbilicus, where the liver arises, is also the anlage of the *duodenum*. The liver can thus be described as a duodenal gland on account of its partial origin from the epithelium of the duodenum. There are numerous other duodenal glands, some of which remain so small throughout life that they stay within the intestinal wall. On the other hand, the *pancreas* is a medium-sized visceral organ that arises from the duodenal epithelium in a similar manner to other duodenal glands, but grows beyond the intestinal wall.

The correlated increase in the volume of the heart and the liver has a unique constructive relationship to the growing vertebral column, which lies close to both these organs. While the curved vertebral column is extending to form a straighter rod-like structure, the heart and liver are enlarging to become more globular (Fig. 7.19). A "vacuum" is therefore tending to form between the growing heart, liver, and vertebral column, in the so-called paravertebral *heart–liver angle*. The suction field here provides the occasion in normal phenogenesis for the development of the lung.

Lungs: One of the many significant consequences of the descent of the heart is the development of the *respiratory tract*. As before, we find

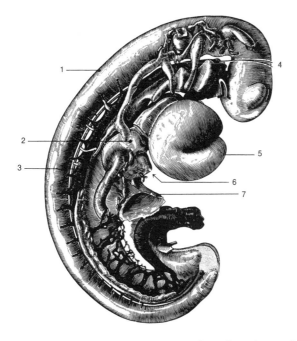

Fig. 7.16. Serial section reconstruction of a 2.57 mm long human embryo (Carnegie Stage 12). Ectoderm removed. Survey of vascular organs. 1) spinal cord, 2) site of junction of great veins with the heart (sinus venosus), 3) aorta, 4) junction of 1st visceral arch artery with dorsal aorta and the connective tissue "bridle" of brain at the nasal placode, 5) heart loop (ventricle), 6) liver, 7) cut edge of yolk sac.

that the differentiation of the respiratory tract is also simply a local modification of the total phenomenon of phenogenesis. The wall of the foregut is closely flanked by and bound to the great venous channels from the head region (right and left superior cardinal veins). The

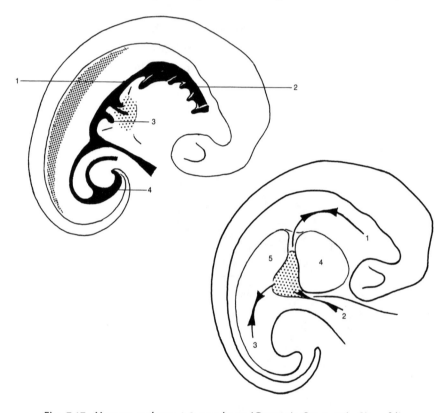

Fig. 7.17. Human embryo 4.2 mm long (Carnegie Stage 13). Site of liver development loosely stippled. Site of embryonic excretory apparatus (mesonephros) finely stippled. Endoderm black. 1) esophagus, 2) pharyngeal part of gut with laterally directed pharyngeal pouches, 3) gall bladder, 4) anlage of urinary bladder.

Fig. 7.18. Schematic drawing of the same 4.2 mm long embryo as in Fig. 7.17 based on serial section reconstruction. Liver stippled. Converging double arrows: anchoring of great veins in liver and their restraining functions. 1) root of head vein (superior cardinal vein), 2) root of umbilical veins, 3) root of intestinal (portal) vein, 4) pericardium, 5) peritoneum.

descending heart draws off a segment of the foregut wall into the territory of cardiac development well before the end of the 1st month of development. Part of the foregut wall is therefore carried along with the descent of the heart and the accompanying veins and so takes part in the entire process of descent. The "drawn-off" portion of foregut wall is the anlage of the respiratory tract, which initially is just a small cul-de-sac (diverticulum) of the mucous membrane lining the gut. Therefore, the respiratory tract lies ventral to the esophagus, that is, in front of it (Figs. 7. 17, 7.23).

When the embryo is 10 mm long, the heart and liver have increased in girth and the space between the heart–liver mass and the vertebral

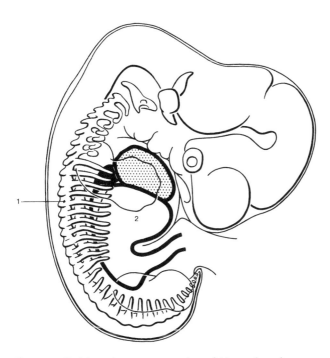

Fig. 7.19. Serial section reconstruction of 10 mm long human embryo (Carnegie Stage 17) showing lung development in heart–liver angle adjacent to vertebral column. Skeleton outlined, heart stippled, lung black, liver white. 1) 5th thoracic vertebral body, 2) liver. The thick black lines indicate the serosa of the pericardial and abdominal sacs.

column has enlarged just beneath the stretched skin of the lateral body wall. A suction field arises in this space (Figs. 7.19, 7.20). The delicate endoderm of the anlage of the respiratory tract, together with its stroma, invaginates into this suction space. The in-growing tissue is the so-called *lung bud* (anlage of lung). This in-growth occurs all the more easily because the endoderm has a high capacity for surface growth and because the space between the heart–liver mass, the vertebral column, and the body wall has been part of the fluid-filled body sac (coelom) since the time of early development. As the space between these three

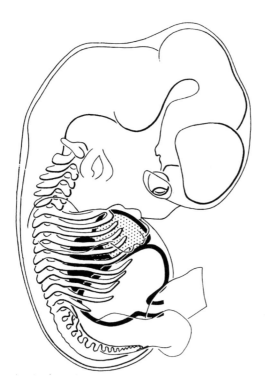

Fig. 7.20. Serial section reconstruction of 17.5 mm long human embryo (Carnegie Stage 20) showing location of lung development. Thick black lines indicate the serosa of the pericardial and abdominal sacs. Heart stippled, lung black, liver white. The thorax has enlarged in association with the growth elongation of the vertebral column and ribs.

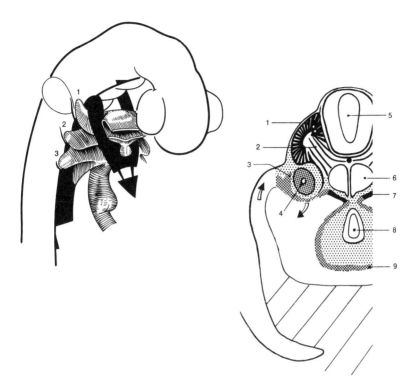

Fig. 7.21. Segment from serial section reconstruction of 3.4 mm long embryo: head–neck region (ectoderm removed) viewed obliquely from right front aspect. Foregut endoderm in head and neck is hatched. Pair of 1st visceral arch arteries black. 1, 2, 3 indicate 1st, 2nd, and 3rd pharyngeal pouches. Broad arrow denotes descent of heart. The angle between the left and right 1st visceral arch artery decreases during the constant pull of heart descent, and the endoderm at the floor of the pharynx is "worked off" to form the anlage of the thyroid gland (seen above the arrow).

Fig. 7.22. Part of cross-section of 3.4 mm long human embryo about 27 days old (refer to Fig. 2.40) showing site of development of excretory apparatus and genital glands at posterior wall of abdomen. Arrows indicate rotation of peritoneal fold (urogenital ridge). 1) ectoderm with dermatome of somite below, 2) sclerotome of somite, 3) anlage of mesonephric (Wolffian) duct in floor of groove on lateral side of urogenital ridge, 4) inner lumen of the embryonic excretory apparatus (mesonephros), 5) neurocoele, 6) aorta, 7) anlage of genital gland, 8) intestinal tube, 9) peritoneum. Inner tissue stroma loosely stippled. Ventral body wall hatched.

components increases, so too does the volume of the body sac (so-called pleuroperitoneal canal). The enlargement of this space always precedes the gentle but expansive surface growth of lung bud endoderm (Fig. 7.25). As the heart, liver, and lungs continue to grow, the initial single-chambered body sac gradually becomes subdivided into the pericardial sac, the peritoneal sac, and two pleural sacs.

During fetal development, the space available for lung growth increases still further by lateral expansion of the thorax. The lungs are

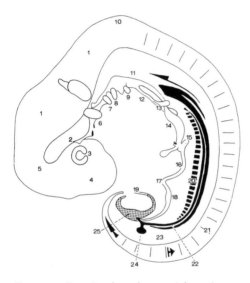

Fig. 7.23. Drawing based on serial section reconstruction of 6.3 mm long human embryo (about 30 days old, Carnegie Stage 14). Site of definitive kidney formation in loosening angle (23) between the mesonephric (Wolffian) duct (22) that remains short and the elongating spinal cord (21). Upper half arrow: restraining function of great vessels. Lower diverging arrows: growth elongation of spinal cord. 1) hindbrain, 2) hypophysis (so-called pituitary gland), 3) eye, 4) forebrain, 5) midbrain, 6–9) 1st to 4th pharyngeal pouches, 10) cervical flexure at junction of medulla oblongata and spinal cord, 11) (below) pharynx–larynx region, 12) (above) trachea, 13) lung buds, 14) stomach, 15) pancreas opposite the gall bladder, 16) duodenum, 17) apex of primary gut loop, 18) large intestine (cecum), 19) allantois, 20) mesonephros, 21) spinal cord with adjacent somites, 22) mesonephric (Wolffian) duct, 23) suction space (see text), 24) anlage of kidney (metanephros), 25) anlage of urinary bladder.

therefore sucked into the growing thorax as pulmonary lobes, similar to their subsequent movements in breathing; this is *growth inspiration*. The increasing surface area of the endoderm of the lung is associated with an increased release of fluid, due to the diathelial function of a limiting tissue. This fluid ebbs upwards from the lung into the amniotic sac as a kind of *growth expiration*.

During the growth expansion of the chest, the *ribs* become tilted somewhat inferiorly as a consequence of the descent of the viscera, and become disengaged from the vertebral column through growth "fracturing." This "fracturing" leads to the formation of joints between the vertebrae and the ribs (costovertebral and costotransverse joints).

The very development of the respiratory tract and the lung is therefore a remarkably differentiated beginning of the subsequent activity

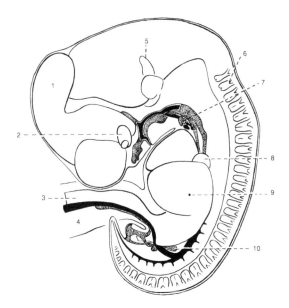

Fig. 7.24. Serial section reconstruction of 10 mm long human embryo (Carnegie Stage 17). Endoderm stippled, arteries black. 1) midbrain, 2) eye, 3) peritoneal lining of coelom of the umbilical cord, 4) umbilical cord, 5) otic vesicle (endolymphatic sac part), 6) 2nd spinal ganglion, 7) thymus with larynx (stippled) above and thyroid (black) below, 8) lung, 9) liver bulge, 10) calyx of kidney.

we call breathing. Strictly speaking, it is incorrect to talk of the "first" inspiration after birth. Breathing movements, by which air is sucked into and expelled from the lungs, are late consequences of the most complicated processes that were established and regulated long before birth.

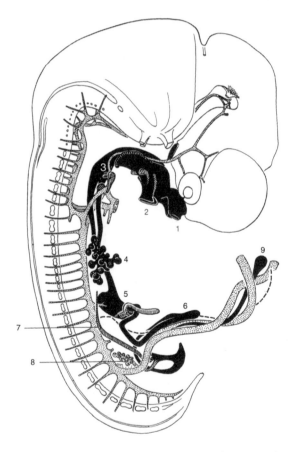

Fig. 7.25. Serial section reconstruction of 13.5 mm long embryo (Carnegie Stage 18). Arteries stippled, viscera black. Dashed line: yolk sac (omphalo-enteric) artery. 1) nostril, 2) opening to mouth, 3) larynx, 4) lung, 5) stomach, 6) primary gut loop, 7) mesonephric (Wolffian) duct, 8) renal calices, 9) allantois (black) with paired umbilical arteries (stippled). Note the similarity between lung and kidney growth in their respective suction fields.

The thyroid as a paradigm for all endocrine glands: In the head region of a 2–3 mm long embryo, we observe that the paired aortic arteries adjacent to the neural tube are flexed in association with the longitudinal bending of the entire neural tube (Fig. 7.16). Pairs of vascular bridges (**anastomoses**) arise between the outflow path of the heart and the dorsal aortae in the head region. These bridges constitute the so-called vascular cage of the foregut (see Fig. 2.38a). The first pair of anastomosing vessels branch from the single ventral vessel ascending from the heart directly under the entrance to the mouth. Initially the angle between these branches is relatively obtuse; however, it becomes increasingly acute as the heart descends. The endodermal floor of the foregut is normally pinched within this decreasing angle to form a small diverticulum (Fig. 7.21). The process is similar to the origin of the respiratory tract. The tiny, squeezed diverticular formation is the anlage of the subsequent vital organ, the *thyroid gland*. As the heart descends in the biomechanical metabolic fields of this region, the diverticulum soon becomes separated from its parent tissue. Similar epithelial separations can also be demonstrated at other sites under analogous circumstances. The differentiation of the thyroid gland is a typical example of a gland that loses its excretory duct. Such glands without excretory ducts are called incretory or **endocrine glands**. Once contact with the original limiting tissue is lost, the disconnected tissue becomes an organ that forms a close association with blood vessels. During the ascent of the head, the anlage of the thyroid is anchored in the mid-region of the neck where it remains, lying close to the thyroid cartilage of the voice-box (larynx). The now so-called thyroid gland is palpable in this region. In disease, the thyroid gland may enlarge to become a goiter.

Larynx

In the mouth–throat region, the entire foregut is compressed between the enlarging brain above and the beating heart below. Inferior to this region, the direction of compression changes rather abruptly by 90 degrees, that is, the foregut becomes squeezed from lateral to medial due to the presence of adjacent, taut blood vessels in the inner tissue. The voicebox or *larynx* (**glottis**) arises in the neck region at the site of this lateromedial compression of the foregut (Figs. 7.24, 7.25). The larynx represents a hybrid structure that exhibits characteristics typical of the organs of movement as well as the viscera. The skeletal elements and muscles of the larynx are like those of a limb and its **mucosa** (mucous membrane from endoderm) is typical of the viscera. The skeletal elements include the thyroid cartilage, cricoid cartilage, and vocal cords, as well as other smaller structures. Although the larynx appears to be a most complicated organ, its anatomy is the outcome of a straightforward developmental sequence, as follows.

Densation fields arise in the depths of the 3rd and 4th visceral arches, just as in the 1st and 2nd arches. Regional comparisons show us that, from the viewpoint of developmental dynamics, these fields are no different from any other foci of skeletonization. The foci have subtly different morphologies according to the local shape of the epithelium (in this case, a mucous membrane) in the vicinity of the cervical viscera. The so-called *vocal cord apparatus* of the larynx is the result of specific epithelial folds in the mucosa lining the wall of the foregut.

On the one hand the developmental movements of the larynx are related to the descent of the cervical viscera and, on the other, to the piston-like growth of the embryonic cartilaginous skeleton of the lower jaw. In correlation with both of these growth functions, particularly that of the cartilaginous mandible, the superior part of the larynx is displaced relatively far toward the front of the neck (ventrally). The strongest cartilage of the larynx, the thyroid cartilage, is palpable under the skin of the neck as a skeletal structure that moves on swallowing ("Adam's apple"). Once again, all these differentiations are necessary consequences of developmental dynamics and are not to be interpreted, say, as historical carryovers.

Lymph Vessels and Lymph Nodes

One of the consequences of the further development and elaboration of blood vessels is the formation of a fluid that is expressed from the vessels with increasing blood pressure. *Lymph* is thus a type of fluid that initially does not serve for nutrition but is rather a by-product of metabolism and growth. Lymph congests in the interstices of inner tissue and *lymphatics* (lymph vessels) arise as soon as these interstices communicate with one another. Lymphatics enlarge and eventually drain into large veins.

The formation of lymphatics is a prerequisite for the emergence of *lymph nodes*. If one compares the various places in the body where lymph nodes are found, then it can be established that they normally arise wherever the flow of lymph is obstructed at sites of kinking in the growing embryo. Such regions of lymph congestion arise wherever there are "bottlenecks" in the body; for example, in the region between the head and neck where the neck flexes, and at the creases where the limbs join the trunk in the vicinity of the armpit (*axilla*) and the groin (*inguinal region*). Initially a type of highly cellular, diffuse lymphoid tissue congregates in these congested zones in the meshwork of the interconnecting lymphatics. Lymph nodes arise from this diffuse tissue. Lymphatic tissue is comparable to a chemical filter where substances that are foreign to the body and poisonous are purified. Whereas blood vessels are serving nutritional requirements, the body uses lymphatic structures for its own detoxification. Lymphatics and lymph nodes become especially significant in cases of infection, for example, in septicemia.

Excretory Apparatus Including Kidneys

The very formation of the *excretory apparatus* is also the beginning of a specific activity that is a prerequisite for its adult function (functional development). With the narrowing of the embryo that takes place in the 4th week, an immense longitudinal fold of peritoneum forms on either side at the posterior wall of the abdomen. The fold is called the urogenital ridge; it contains the embryonic kidney (**mesonephros**; Figs. 7.17, 7.22, 7.31). The inner tissue of the fold has the form of a long spindle and the fold is bounded laterally by a groove in the peritoneum. The unique differentiation of this longitudinal fold is coupled with a slight degree of outward rotation about its longitudinal (spindle) axis. The rotation occurs because the adjacent, large paired aortic vessels are straightening and moving slightly away from the spinal cord, dragging the attached peritoneum along with them. Meanwhile the peritoneum in the vicinity of the lateral groove described above does not move forward. This combination forces an outwards rotation of the fold (Fig. 7.22).

The inner tissue (stroma) of the peritoneal fold becomes cylindrically oriented and then, even in the 1st month, hollow. A spindle-shaped vesicle or cyst therefore arises in the fold. This vesicle in turn becomes subdivided lengthwise into a row of very much shorter vesicles, the so-called embryonic urinary vesicles (*mesonephric glomeruli*). The epithelium of the lateral groove also becomes hollow and so gives rise to a longitudinal tube known as the *mesonephric duct* (Wolffian duct). Each embryonic urinary vesicle elongates and twists forming an S-shaped *mesonephric tubule* that comes into contact with the epithelium of the Wolffian duct. The site where each tubule contacts the mesonephric duct is a corrosion field; cell death here leads to a communication between the fluid in the embryonic urinary tubule and the lumen of the duct.

As the liver enlarges, the entire superior part of the embryonic kidney is compressed and disintegrates. Only the inferior end of the Wolffian duct has the spatial opportunity for further development, and this is where the definitive *kidney* (**metanephros**) arises. During its functional development, the definitive kidney exhibits the following developmental movements.

At the start of the 2nd month, in say a 6 mm long embryo, the Wolffian duct is bent just superior to where it drains into the anlage of the urinary bladder. At this bend, the wedge-shaped epithelial cells of the duct diverge quite markedly toward the periphery of the duct. This wedge epithelium thus attains a unique spatial opportunity to sprout from the duct wall. The precondition for this process is a suction field to which the wedge epithelium adapts by sprouting. The suction field arises at the inferior end of the body where a highly localized "loosening space" is starting to form between the elongating neural tube, which is moving dorsally, and the mesonephric duct, which is remaining relatively ventral (Fig. 7.23). The wedge-shaped epithelial cells grow into this space to form the anlage of the definitive kidney, an initially blind-ended diverticulum (metanephric bud). This single diverticulum already has the basic form and architecture of a kidney with superior and inferior poles and a "capsule" composed of stromal cells (metanephric blastema). The cells at the poles are the most wedge-shaped of the whole diverticulum and on account of their higher growth rate are responsible for further enlargement of the diverticulum. Additional blind-ended diverticula therefore arise at each pole of the metanephric bud. These diverticula continue to branch so that a highly ramifying gland-like structure gradually arises (Fig. 7.25). As with the lung, the development of the definitive kidney is a beautiful example of **encapsis** or the repetition of similar forms at successive orders of magnitude.

Here we will go no further into the more complex architecture of the kidney; it suffices to establish that the young kidney is already functioning as an excretory organ. During its growth, the definitive kidney removes nutrients from its surrounding stroma and secretes highly fluid by-products into the widening lumen of the excretory apparatus. In this way, an excretory activity by the kidney is established in quite early stages. The early "pre-urine" flows into the anlage of the urinary bladder and then into the allantois before entering the amniotic sac and umbilical arteries, initially by diffusion.

The Gut

The gut also displays a phenogenesis with dynamic components. One of its early characteristics is so-called *gut rotation*, which is a growth rotation of the intestine about its long axis. At this time, the apex of the intestinal loop has already lost its connection with the yolk sac. Gut rotation occurs in the following manner. At the end of the 1st month, the alignment of the plane of the intestinal tube as seen from the front (ventrally) appears to be almost straight (sagittal). However, when observed more closely the plane containing the gut is found to be slightly twisted. Superior to the level of the umbilicus, the gut tube lies a little to the left (where it is the anlage of the stomach) and in the mid-region of the umbilicus, a little to the right (duodenum anlage). Inferior to the umbilicus, the gut lies again more to the left (Fig. 7.26).

This last segment gradually forms numerous secondary loops and develops into most of the coiled *small and large intestine*.

The so-called *primary gut loop* in the inferior region of the umbilicus of an embryo about 6 mm long is shown in Figure 7.27. The impetus for gut rotation is cell division in the *intestinal endoderm*: successive waves of epithelial cell division spiral along the gut tube. As the gut grows, the

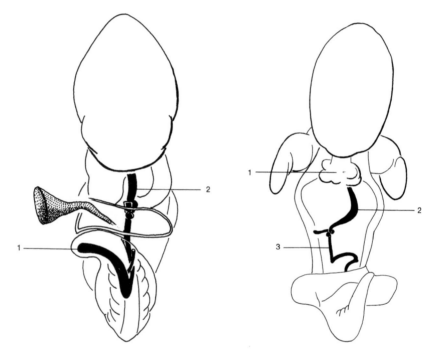

Fig. 7.26. Serial section reconstruction of 4.2 mm long human embryo (Carnegie Stage 13) showing development of visceral tube (black) as yet hardly twisted. Yolk sac cut off at yolk sac stalk (stippled). Cut edge of umbilical cord is double-outlined. 1) hindgut, 2) stomach.

Fig. 7.27. Serial section reconstruction of 6.3 mm long human embryo (Carnegie Stage 14) showing the growing intestinal tube twisted between its fixed points in the superior and inferior body wall. 1) trachea with pulmonary lobes (three on right, two on left), 2) stomach and upper part of duodenum with gall bladder, 3) duodenum at back of abdominal wall. The primary gut loop extends inferior to the duodenum into the umbilical cord as far as the ventral body wall (secondary loops have not yet formed).

apex of the primary loop moves in a circular arc, in a counterclockwise direction when viewed from the front. As the anlage of the appendix lies near the apex of the primary gut loop, the appendix shifts with the rotation. In cases of incomplete rotatory growth movements, the appendix might finally arrive, for example, in the vicinity of the liver on either the right or left side of the body. Thus in certain cases inflammation of the appendix may be very difficult to diagnose.

The lumen of the superior part of the gut tube is initially wider than that of the inferior. The superior part becomes filled with secretions from the large intestinal glands, particularly the liver and pancreas. Conversely, the lumen of the inferior part of the gut tube remains tiny and the perimeter of the tube very small. Thus the entire inferior part of the gut loop is initially less conspicuous than the superior part. As the inferior segment elongates, numerous secondary gut loops arise (Figs. 7.28, 7.29). Initially these secondary loops remain within the root of the umbilical cord in the so-called umbilical coelom (Chapter 2). This

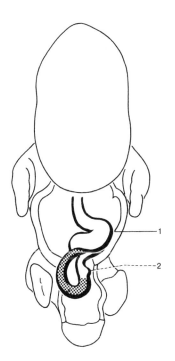

Fig. 7.28. Drawing based on a serial section reconstruction of 13 mm long human embryo (Carnegie Stage 17). The umbilical cord (below) has been opened in the model to expose the umbilical coelom containing the primary gut loop (stippled). 1) stomach, 2) cecum ("blind gut") indicating site where appendix arises.

location of the gut is frequently described as the *physiological umbilical herniation*, even though the tube here is simply growing *in situ* (Fig. 7.28). It is only after the ventral wall of the abdomen extends far in front of the enlarging liver that the "herniation" disappears. In other words, the abdomen grows to enclose the intestines; a true repositioning of the gut loops (in the sense of a posterior displacement of a hernia) does not occur.

The lumen of the gut widens with the development of numerous secondary loops. The motor driving this re-arrangement is still the epithelium of the mucous membrane (i.e., the intestinal endoderm). By means of cell division, gut endoderm stretches the inner tissue and its vascular network into a cylindrical configuration. As a result of this activity, ring musculature differentiates in circular dilation fields. In distinction to skeletal muscles arising in association with the piston-like growth of cartilage, we could designate the developing ring musculature of the intestine as *epithelial musculature* on account of its formative affinity to the endoderm.[17] The fibers of this ring musculature gradually impede further enlargement of the caliber of the endodermal part of the gut tube. This restriction forces the gut to increase in length and this in turn leads to the origin of a new dilation field external to the ring musculature: this is the site where the longitudinal musculature develops. Its growth dilation also is therefore secondary: the longitudinal musculature lies peripheral and the ring musculature more to the inside. The formation of the entire gut musculature is probably a rhythmical process proceeding from superior regions to inferior; this is possibly a precondition for the subsequent **peristaltic** activity of the gut. An alternating rhythmical process of circular dilation and longitudinal dilation may explain why the directions of muscle fibers in the wall of the adult intestine are pitched at slight angles to transverse and longitudinal directions, that is, why they spiral and are not exactly circular or longitudinal.

High and low relief formations normally arise in the lining epithelium as soon as the resistance of the ring musculature to stretching begins to impede further enlargement of the lumen (Fig. 7.30). The changing relief is a sign of further intensive surface growth of the intestinal endoderm. **Villi** arise from the high formations, and intestinal glands (crypts) from the low ones. The *intestinal villi*, covered by diverging wedge-shaped cells, increase the contact area between the intestinal mucosa and the contents of the intestine. As a result of the more vigorous surface growth of the gut (as opposed to volume growth), a pressure gradient develops from the surface of the villi toward their stroma: this gradient can be readily diagnosed by its morphological signs. Corresponding to

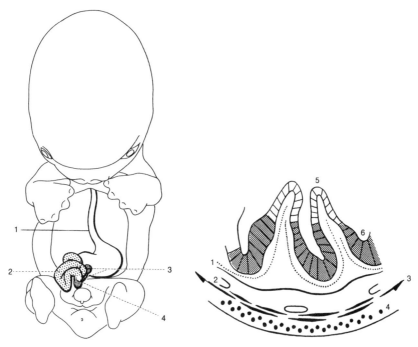

Fig. 7.29. Human embryo 20 mm long (Carnegie Stage 20). 1) esophagus near junction with stomach, 2) secondary intestinal loops, 3) large intestine (colon), 4) appendix.

Fig. 7.30. Transverse section of small intestine from a human fetus about 50 mm long. Apices of intestinal villi lightly hatched, crypts hatched in dark-gray. Black line at 1) interface between mucosa and submucosa. Dotted line at 1) site of future muscle layer of mucosa (muscularis mucosae) dilated by elongation of villi. 2) blood vessels in submucosa, 3) growth extension (dilation) of ring musculature (diverging half arrows), 4) longitudinal musculature cut in cross-section, 5) villus, 6) crypt.

this pressure gradient, the apices of the villi gain the capacity to absorb the intestinal contents. On the other hand, the converging wedge-shaped cells located in the *intestinal crypts* between the villi obtain a broad contact surface with the inner tissue. Here the cells remove raw materials from the vascularized stroma (i.e., stroma nourished by blood vessels) and produce glandular secretions.

Genital Glands (Gonads)

It is possible to demonstrate the presence of *germ cells* in the tissues of the inferior half of the human embryo as early as the 3rd week of development. These germ cells do not migrate of their own accord but multiply *in situ* as the embryo grows. By the 4th week of development, when the embryo is about 2 mm long, a narrow strip of peritoneum and germ cells forms a covering over each aorta. This strip is the anlage of a genital gland (**gonad**; Fig. 7.22); its cells are nourished directly from the aorta. The significance of this location is that the genital glands are protected from the effects of external influences and so initially undergo no differentiating growth, but rather preserve their innate individuality. Both the gonad and the embryonic excretory apparatus (mesonephros) grow in the longitudinal fold called the urogenital ridge, with the gonad elongating on the medial aspect of the embryonic kidney (Fig. 7.31).

It is only in the 3rd month that the gonads display obvious significant differences in male and female fetuses (Figs. 7.32, 7.34). The soft, elongated female genital gland (*ovary*) lies with its inferior border flat against the lateral wall of the small (minor) pelvis. The more cellular and more globular, encapsulated male genital gland (*testis*) finds no room to grow within the minor pelvis. It soon emerges into the large (major) pelvis along the line of least resistance and lies tilted laterally on the wings of the pelvis (iliac fossa). From there the testis glides through a weak region in the ventral abdominal wall, through the inguinal canal and into the scrotal sac (Figs. 7.33, 7.35). The testis does not "migrate" of its own accord. Rather, each testis follows a guidance structure (gubernaculum testis), which is the ventral segment of a much longer restraining band of stretched inner tissue that passes in

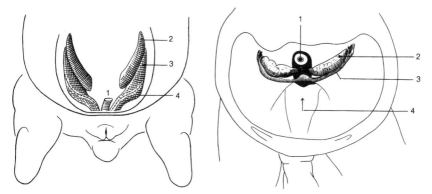

Fig. 7.31. Excretory apparatus and genital glands (gonads) in 27.2 mm long embryo. 1) hindgut (rectum), 2) genital gland (ovary), 3) embryonic excretory apparatus (mesonephros), 4) region of mesonephric (Wolffian) duct. The paramesonephric (Müllerian) duct (anlage of uterine tube, etc.) is indicated by the converging dashed lines.

Fig. 7.32. Pelvic organs in a female fetus about 100 mm long. 1) rectum in cross-section, 2) ovary, 3) uterine tube, 4) (above arrow) common junction of both uterine tubes to form uterus.

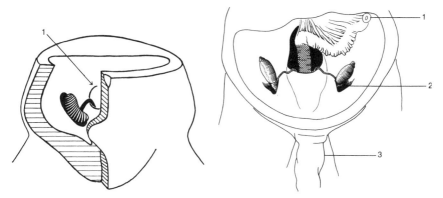

Fig. 7.33. Start of descent of testis in 55 mm long male fetus. Pelvis partly opened. Testis dark closely hatched; epididymis and seminal tube (ductus deferens) black. 1) entrance to minor pelvis.

Fig. 7.34. Pelvic organs of a male fetus about 115 mm long. 1) sigmoid colon, 2) entrance to inguinal canal with testis and ductus deferens (from mesonephric duct) lying just superiorly. The mesonephric (Wolffian) duct does not develop to any significant extent in the female fetus. 3) umbilical cord.

an arc out of the pelvis and terminates dorsally in the vicinity of the end of the vertebral column (testococcygeal ligament). As the total length of the testococcygeal ligament scarcely changes with growth (particularly during the piston-like growth of the cartilaginous pelvis and vertebral column), the testis shifts relative to the pelvis. In cases of abnormal growth, the displacement of the testis may fail so that it is retained in the abdominopelvic sac; alternatively, the testis may slip past the scrotum to the vicinity of the anus (ectopic perineal testis).

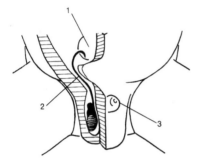

Fig. 7.35. Descended testis in a fetus about 240 mm long. 1) entrance into the minor pelvis, 2) ductus deferens in inguinal canal, 3) genital tubercle (glans penis).

Chapter 8

Human Behavior Patterns Are Initiated through Ontogeny

If one were asked to state the two characteristics most typical of humans, one would probably say: speech and the upright gait. Both speech and the upright gait involve a unique pattern of behavior, and each is preconditioned by a unique human ontogeny. Both are typical behavioral manifestations of human **cerebralization**. Although we still do not understand how human psychological behavior is related to the physiological activity of the brain, we can be certain of one fact, namely that ontogeny and especially the ontogeny of the early embryo is a prelude for all subsequent human behavior. Both speech and the upright gait are the result of a specifically human ontogenetic process; we should not attempt to interpret them as evolutionary modifications of the behavior of our ancestors. The biologist Joachim Illies (1925–1982) captured the irony of such evolutionary ideology in the following verse:

In order to fashion a human being,

Nature—as Dr. Brainstorm claimed—

Had simply to chase the monkey

Right away from the jungle tree;

Because once driven out onto the plains

No other option for monkey remained.

It dawned on monkey from that hour forth,

No more bananas were at his mouth;

He must, if he wanted his life to extend,

Erect his body on two hind legs,

Thereby leaving both hands free

To pick at fruits, do other feats.

The hands, outstretching, pears to obtain

Causes to grow within his brain,

(To safely keep his new-found skills)

From year to year, new grooves and folds,

Until the skull swells to its limit,

One thousand grams contained within it.[18]

◆ ◆ ◆

The Upright Gait

A striking characteristic associated with the upright gait is the architecture of the skeleton. Morphological comparison of human and animal skeletons shows that in human adults, the inferior portion of the pelvis (the so-called minor pelvis) is inclined about 30 degrees relative to the frontal plane of the superior (major) pelvis. The adult pelvis therefore has a distinct bend in the transition region between the major and minor pelvis. The transverse axis of this region passes through the hip joint. Thus in humans, most of the mobile portion of the vertebral column is located vertically over the hip joint (see Fig. 6.1). This anatomy, which is quite different from that found in quadrupeds, is essential for the upright gait and arises hand-in-hand with the following ontogenetic growth movements.

In human cerebralization, the growing forebrain is a substantial sink for nutrients; this sink provokes the development of a massive heart to supply the brain. The size of the embryonic heart demands a powerful peripheral blood circulation. This, in turn, enables an intensive growth of the liver. The growth of the heart is usually correlated with the growth of the liver. As described in Chapter 7, the intervening diaphragm is compressed into a thin sheet between the heart and liver and therefore becomes tendinous. The liver is always fixed to the inferior surface of the central tendon of the diaphragm. With growth, the heart–liver mass and the brain move apart. The entire visceral tract accompanies the descent of the more dominant heart–liver mass (**descensus**). Relative to the diaphragm, the brain and spinal cord are ascending at the same time. This ascent is associated with further growth of the brain particularly of the cerebral hemispheres, which are growing eccentrically both upwards (superiorly) and backwards (occipitally). The enlargement of the occiput is the precondition for an enormous widening of the angle that the head subtends at the nape of the neck. The powerful dilation fields for the neck musculature arise in this angle during the ascent of the brain (Fig. 6.15). Contraction of this musculature erects the head and this extension conditions the development of the upright gait.

The emergence of the upright gait is therefore a result of cerebralization and thus, of the entire ontogenetic process. Other animals do not manifest a similar dominance of brain development and therefore do not develop the upright gait.

The Somatic Basis for Our Awareness of the Body As an Instrument

We already have prescience of our own bodies as a result of our prenatal development. This knowledge manifests itself unmistakably in our many gestures, even though we are rarely conscious of it. In considering this, it is impossible to separate the psychic from the somatic.

Generally, seldom emphasized is the extent to which the very growth of the human brain is a preparation for its subsequent importance. The early ontogeny of the brain is decisive in laying the foundations for all subsequent patterns of human behavior. The conventional view is that the cerebral hemispheres grow symmetrically. Reconstructions of the brains of embryos and fetuses have demonstrated the following remarkable phenomenon concerning the cerebral hemispheres. Normally each hemisphere grows in reciprocal harmony with the other, as follows.

As the volume of one hemisphere (say the left) is increasing through surface growth, then the opposite (right) hemisphere will be pressed more firmly against its dura mater and thus impeded in its volume growth. This impediment causes a smaller increase in the surface area of the right hemisphere and, consequently, a thickening of its wall. In turn, this thickening of the cerebral wall demands additional nourishment. The influx of nutrients to the external, highly vascular layer (pia mater and arachnoidea) of the right hemisphere therefore increases. As a consequence of wall thickening, the nutrient influx to the right hemisphere now exceeds that to the left. This influx favors anew the surface growth of the wall of the right hemisphere, against the yielding resistance of the dural capsule. The left hemisphere, in its turn, now becomes pressed more closely to its dura. In other words, what appears to be a symmetrical growth of the cerebral hemispheres is really an oscillating process between the right and left sides of the head. Long before birth, the two hemispheres are mutually influencing each other and therefore represent a functional whole. This alternating and reciprocal growth process ushers in the subsequent reciprocal transmission of nerve impulses from one hemisphere to the other. When we say we are "tossing a problem back and forth," we are applying a psychosomatic description to an activity that represents an oscillation from one side of the brain to the other.

An intuitive appreciation of the decisive role of the brain in every sphere of human activity has existed from earliest times, even though the precise relationship between the brain and the head and, in particular, between the brain and the face was quite unknown. Storytellers, painters, and sculptors have long represented human facial expressions without the slightest concept of their physiological relationships to brain activity. Nevertheless, the portrait always embodied an inner activity, as well as being a representation of superficial features; the portrait mirrored the whole personality. In all cultures of the world, the head has always been adorned with jewelry, fabrics, or by particular hairstyles. In early cultures, the crown symbolized an authority endowed with spiritual significance. All these "descriptions" could be classed as a pre-scientific morphology that is just as valid as our scientific morphology. This pre-scientific morphology has existed for centuries, but the study of its history has barely begun.

The psychiatrist Ernst Kretschmer (1888–1964) distinguished empirically three characteristic types of body build (*somatotype*): the "asthenic" or "leptosomic" (slender build), the "pyknic" (compact, heavy build), and the "athletic" (with well-developed muscles). Similarly, the psychologist William H. Sheldon (1898–1977) developed the idea that the somatotype could be analyzed by standardized photographs in terms of three continuously variable qualities of body build: "ectomorphy" (thin physique), "endomorphy" (rounded physique), and "mesomorphy" (athletic physique). Broadly speaking, both of these schemes sought a possible relationship between the somatotype and psychological characteristics, that is, between body build and temperament. Our theory of ontogenetic development allows us to take a small step in this field as well.

We are able to detect varying relationships between the surface area of the human embryo and its volume quite early in the development of different embryos. Ontogenetic development may lead, say, to a human being with a more rounded body build, having a smaller surface area relative to body volume. On the other hand, a more "elongated" individual may mature with a larger surface area relative to body volume. Such individuals often display a tendency for sharpened reactions, as can be demonstrated by graphology. Humans who have developed asthenically to have a lean and extremely "de-rounded" body—so-called "schizothymics" or "cerebrotonic ectomorphs"—frequently display a disposition, not only somatic, but also psychological, toward greater sensitivity and more irritable behavior. Their thought processes may go more easily to extremes. In contrast, rotund people tend to think less "vertically," feel themselves less exposed, and, by way of comparison, are often more easy-going and good-natured. Their conscious self-preservation seems much less endangered. They are truly more self-contained and consequently responsive, with little sense of risk, on many fronts. In comparison to others, they may be more approachable and sociable.

The innate somatotype, being the result of prenatal ontogeny, cannot be varied at will and is therefore of great importance for the constitution of the body as a whole and for individual behavior. In contrast, we have the arbitrary, fluctuating, and short-lived changes in body proportions, which we call *gestures*. The proportions of a particular somatotype are the foundation for the transient changes in body shape that we designate as gestures.

For example, when we move our head slowly from side to side, we sense that it can be easily set in motion, almost like a free instrument poised in a state of labile equilibrium. We can incline the head just as easily to one side as to the other. As a "work-tool," the head can be used for transmitting the message: "It could be like this, or it could be like that." As an instrument of communication, the position of the head is signifying: "Look here, things are still a bit uncertain. The matter is still in doubt; I am of two minds about it."

Any person who has a body like the one making such gestures can, by imitating these gestures, put him- or herself into a state of uncertainty and so perceive instantly the thoughts of the correspondent.

The early development of such body language is a prerequisite for every act of communication, and therefore for the acquisition of all knowledge. How limited our experiences would be if we could not use our bodies for communication! By means of gestures, we are constantly reshaping our own body. In so doing, we are able to use our body to decode the gestures of others. If we could not use our body as an instrument comprehensible to ourselves, then there would be no "self-understanding" of gestures; gestures would be incomprehensible.

If we shrug our shoulders, we are putting our body in a particular configuration so that the shoulder joints no longer permit the arms to be used freely for "hand"-ling. Anyone repeating this gesture under similar conditions will immediately experience the feeling: "It can't be done; it's impossible; I don't have a clue." The message of this gesture is so powerful and so "logical" that it transcends any accompanying verbal or written statement.

Hand Movements As Gestures

When we rejoice to see a baby, barely able to stand, reaching out to touch objects, wanting to grasp everything, then we begin to suspect that without continual bodily "handling," the capacity for mental "grasping" could never develop. It is only after we have used our hands to grasp moveable things objectively that we are able to grasp ideas in an intellectual sense. The fact that children born without limbs can acquire the same mental skills does not disprove the argument that we learn mental skills through somatic gestures. Human intellectual capability still exists in spite of congenital anomalies because, in these exceptional circumstances, dexterity can be acquired by special artificial aids. From the school pupils sitting in a class, able to move their hands freely over the desktop, the teacher naturally expects that they have a capacity to grasp ideas. And in their turn, the pupils take it for granted, in more than just a metaphorical sense, that the teacher who stands before the class already under-"stands" the topic.

In any language, the vocabulary of everyday speech testifies as to how much the hand is involved in comprehension. We talk about "hand"-ling problems, having a quick "grasp," a pre-"conceived" idea, or being "percept"-ive.[19] Idioms such as "to embrace an idea," "to draw a conclusion," or "to seize an opportunity" would hardly make any sense from the point of view of what we have "act"-ually done, if we could not unite, at least in our imagination, the expression with a hand movement. Even the meanings of verbs such as "to refer to," "to assume," or "to reduce,"[20] are based on somatic pre-experience and therefore, on a prescience of our own body and not, for example, on something like an innate, purely logical power of the intellect. We learn to use our hands for reasoning by first creating an awareness of the hands as tools. These ideas have been considered in previous behavioral research, but barely in any more depth than what has been presented here.

There is even a psychosomatic basis for mathematics. When we consciously hold our hands a certain distance apart and place them against an object, we can say we are *measuring* it. On the other hand, we can say we are *counting* if we touch the object sequentially first with one hand, then with the other, and so on. Morphological investigations teach us that our hands are participating in the practical experiences of counting, measuring, and calculating. The conscious and repeated use of our hands as meaningful instruments allows us to acquire the somatic prerequisites for mathematical thinking. The somatic gift for calculating was already recognized by the dealers (i.e., the "hand"-lers of goods) in the market places of the oldest cultures. As an instrument, the hand is capable of communicating particularly objective (i.e., object-bound) messages because it has developed as a very well governed part of the body that can be controlled through the eyes.

Historians teach us that, even in early times, sign languages evolved through many stages and only became standardized after long periods of use. And even longer periods of cultural differentiation were required before sign languages became a psychosomatically reliable prerequisite for complex thinking. Human beings have existed on earth for longer than 150,000 years. Even as recently as Antiquity (2000–3000 b.c.), when the use of the hands was generally associated with the experience of soiling them, it seems that the intellectual manipulation[21] of facts[22] was not widely practiced; it was as if mind and body were separated from one another. Only few could write in Antiquity. The functional development of the hand for performing enlightened tasks became significant in western cultures only with the rise of a scientific way of thinking in the 16th Century. Around this time people learned to perform arithmetic multiplication using the pure and simple symbols **1**, **2**, **3**, etc. These "Arabic" numerals were only standardized toward the end of the 16th Century after thousands of years of psychosomatic preparation. These symbols for the cardinal numbers constitute a unique, internationally recognized, work of art: they are the most outstanding illustrations of the human hand that have ever been created. They are accurate sketches of gestures made by a free-moving hand used consciously as an instrument (Fig. 8.1).

The significance of the hand is also relevant in the seemingly insurmountable gulf between the objective way of thinking in the natural sciences and the subjective way of thinking in the humanities. Using the hand consciously as an instrument, we are able to manipulate technically formed objects and think about them (insofar that the thoughts are bound to an object) objectively. Conversely, we address ourselves to a partner in conversation by using gestures of the mouth. Such gestures would never be used for objects, but are reserved for human beings. Thus on the one hand, we have the objectivity of mathematics, and on the other, the individual humanity of language.

As well as the cardinal numbers, the invention of mathematical operations is based upon a somatically conditioned pre-experience. For example, we are able to experience the seizing of an accessible object, say a rod, not only single-handedly, but also with both hands. After seizing the rod, we are able at will to slide both our hands so close together

Figure 8.1. Depiction of "Arabic" numerals with the hands, based partly on gestures for numerals still used today in some oriental markets. Note that the "weaker" numbers 1–5 are usually conveyed by the supinated left hand and "heavier" numbers 6–9 by the pronated right hand (after Blechschmidt, 1966).

that there comes a point between our hands in which our capacity to continue the operation disappears. We can designate this contact as the "nullpoint" or zero between positive and negative values.

The symbols for plus and minus also have a psychosomatic origin. Along a rod that is suitable for gripping, we can slide a hand to the right or to the left gripping the rod at intervals, that is, measuring. In our hands the rod becomes a measuring stick and *plus* signifies: the open hand moving to the right and at a certain point catching hold of the rod with a closing movement. The symbol "+" can be thought of as a sketch of these two successive voluntary movements that are mutually perpendicular. Similarly, *minus* signifies: the hand remaining open, moving both to the left and away from the rod. The symbol "–" is a sketch of such a hand movement.

The farther we stretch out our arm, the heavier it becomes; correspondingly, the larger, more numerous, and more momentous becomes a number in a positive as well as in a negative sense. Therefore, the larger numbers are at the extremes of the scale of "finite"[23] magnitudes. If we flourish our hand (like a conductor before an orchestra) swinging it alternately palm-upwards and palm-downwards, we can represent a concept of infinity. The concept becomes comprehensible through movement, that is, somatically. This gesture too combines mind and matter at the one time. And like others, this gesture may be represented symbolically; in this case by "∞."

The equal sign is another symbol that has mental and somatic prerequisites. When we measure off the two ends of an object between the open palms of our hands, we experience simultaneously the object itself and the intervening distance between our hands. We become aware of two reciprocally identifiable magnitudes: the distance between the ends of the object and the separation of our hands. We can illustrate the equivalence objectively with a sketch of the hands as the double strokes of an equal sign "=." What complications would have arisen in our daily transactions if we had been unable to devise a symbol for equality!

Even the procedures of differentiation and integration of the more "mature" mathematics devised in the 16th Century have psychosomatic foundations. An example is the representation of a perpendicular (Cartesian) coordinate system. As soon as we take the left hand as a zero-point for movements of the right hand (and vice versa), a coordinate system is established. Coordinate systems provide a means, in principle, of working mathematically with variable quantities.

What is the "self"-evident meaning of the praying gesture when people, as they have done since antiquity, raise their hands in supination with palms toward the sky? This gesture clearly expresses a readiness to receive humbly something greater from above. This gesture of reverence is truly a *supinare ad superos*. It was experienced as something palpably different from, for example, the gesture of clasped hands signifying "I feel obligation-bound," or the gesture of the rejecting, outstretched right hand with pronated palm and dorsiflexed wrist signifying "Stop!" Anyone who acts out the gesture of the stiffly outstretched and obliquely raised upper limb with pronated hand has a vivid experience of the thought: "Everything is 'sub'-jugated to my will, dependent on me! Everything is exposed to the grip of my hand."

The psychosomatic experience is totally different for the menacing, the protecting, or the blessing hand. If the arm is outstretched and the hand clenched into a fist with tensing of the muscles of the forearm and hand, then a raw force can be physically perceived. Conversely, the gesture of a slow extension of the pronated hand with slightly bent 4th and 5th fingers indicates a benedictor "just beginning an action, visible to others."

Hands thrust into pockets are also "self"-evident messages; under certain circumstances they may signify: "As you see, my hands are not there. There is nothing I can do, nor can I change anything." Whoever puts only one hand into a pocket is making the statement: "I'll handle the matter in this way and in no other way. Right now, any other way is out of the question." One of the special thrills of the theater is that dramatic art can succeed in transporting us into the realm of somatic experience.

❖　❖　❖

In earlier chapters, we noted relationships between the micropopulations of cellular ensembles and organs on the one hand and macropopulations, in the sense of groups of people (socializations of higher orders), on the other hand. This means that the behavior patterns seen in macropopulations may share common features with the patterns of behavior that we can observe in the living cells of the human organism. If we were unable to react somatically to living processes occurring inside our bodies, then we would be equally unable to perceive relationships in social life and to act accordingly. Without an innate awareness of living bodily processes, higher levels of human endeavor involving vocational differentiation and more complex cultural development could never be attained.

If we were incapable of making gestures with our bodies, then no language, no use of language for thinking, and no social order would have developed. What we call *sociology* is based on the ontogenetic development of a "sociology" of cells and cellular ensembles. Our body provides us with such wealth, far greater than we could ever have anticipated!

In addition to intellectual activity, our somatic development prepares us for emotional or affective impulses. Somatic development is the basis for our capacity to communicate our thoughts and feelings. The person who is laughing is somatically different from someone who is not laughing. The person who is reflecting deeply upon something is, in their metabolism and therefore physically, tuned differently and altered in comparison to someone who is talking, or telephoning, or singing.

All gestures are gestures of the body. We see them and we imitate them. Everyone knows that laughter is infectious. Indeed there are patterns of behavior that are not only transferable, but which, in their transfer, can develop to take on totally different meanings. A large part of what we call "learning" is simply the history of the human race in a continual process of transformation by ever renewing generations. Trained and enlightened, we translate the living experiences of our physical bodies into abstract concepts and logical thoughts. To be able to behave in this way requires that we already have at our disposal a living, pre-scientific knowledge of our own body. For the path to abstraction comes from observation.

Instincts

Psychologists and pediatricians believe that imprinting in early infancy can have repercussions later in life. Many consider this imprinting to be a repetition of phylogenetically earlier behavior patterns. However, such a viewpoint disregards the fact that the human being develops its embryonic performances in a way that is unique to its own ontogeny. It is impossible to comprehend the principles of human behavior patterns by studying only the similarity of human behavior to animal behavior.

It is now possible to prove that all patterns of behavior have embryonic developmental processes as their precursors. This is what leads to the inheritability of so-called instinctive performances. What we call *instincts* are the direct consequences of prenatal developmental events, which are really the prenatal performances of the embryo. Whatever is not already subconsciously initiated and "practiced" by the body during its early development cannot be enacted—either instinctively or consciously—at any later date. As an example, we recall the analysis of the child's suckling reflex (Chapter 5): let us remind ourselves that if the lips of the young embryo had never rolled in as part of an early growth movement, then the newborn child would never be able to suckle instinctively.

The same principle holds true for the clasping reflex: the action of a baby's grasping is a direct consequence of growth grasping. Indeed, preceding growth movements are found for all so-called instinctive reflexes. Congenital instincts are the elaborated reactions of processes that the embryo has already initiated. For example, the baby's attempt to raise itself up is an endeavor to retain and elaborate, under different circumstances, a process that already began in the early embryonic period as an extension of the body during the course of a uniquely human cerebralization.

Therefore, embryonic "pre-experiences" must exist. Such "experiences" have been recognized for thousands of years. One still encounters them today in various guises, as well as in the endless changes of fashion.

Human Dress As an Anatomical Atlas

Fashion, together with the manifold forms of national dress (a once popular manifestation of fashion), is a highly fascinating chapter of knowledge about psychosomatic differentiation. One could even define human dress as the oldest anatomical atlas that has ever been "published."

The collar, for example, emphasizes the narrowness of the neck that arises during the displacement of the heart from this region into the thorax. A collar can therefore be considered natural because the neck is innate, that is, the neck has developed naturally to have its particular form. A necklace mimics the two sternocleidomastoid muscles that, from the 2nd month of development onwards, have been pulling between the occiput and the sternum. A necklace hanging down the back would have no somatic basis. The typically male "Adam's apple" is emphasized by a knot in the necktie. The bridge between the lenses of our glasses recapitulates the most elementary structural formation at the root of the nose, namely the interorbital ligament. In the course of brain development, the inner tissue between the eyes is stretched to become an anchoring band, whose growth resistance during further development holds the eyes together and so determines the typically frontal gaze of the human.

Another example is the buttoning-up of a jacket, which mimics the enclosure of the thorax by the ribs. The mutual growing together of the ribs and the sternum is a pre-experience for buttoning a jacket; buttons on the back are not as "natural."

A bracelet has its somatic foundation laid down already in the 2nd embryonic month, when a connective tissue "bracelet" (**retinaculum**) can be identified in the microscopically small wrist of the embryo.

The halo depicted by artists of the Middle Ages and the tonsure of monks confirm something that dominates human physiognomy, both in the newborn and for a long time before birth, namely the significance of the brain in the uppermost part of the head. It is unthinkable that scientific medicine should claim to have discovered the privileged role of the brain in human mental activity. Here again a vivid pre-experience of the body is evident as a consequence of a lived-through development. That is why somatic dispositions are of major importance for instinctive behavior.

The gesture of the hat cocked on the back of the head with the face free is like any other gesture: it makes immediate sense to us because of its naturalness. At any particular instant such a gesture, on account of its human peculiarity, immediately and spontaneously states: "Let anyone approach me to my face, come right up to me. I feel cocky!"

The hat that is positioned correctly and straight realizes quite a different attitude to life, namely a more fixed and "correct" attitude. Wearing a hat this way expresses a bearing that is obviously not easily shaken. The "gesture" may be accentuated in a "rounded-off" way by wearing a bowler that sits firmly on the head.

The hat that is pulled down over the face conceals it from a partner, creating mystery or stating: "You're not supposed to see what I'm thinking; I'm not going to discuss things with you."

By changing the heels of one's footwear, the length of the stride can be varied by amounts that are scarcely measurable, but yet quite perceptible. "Big wheels" roll along with impressive, long swinging steps. On the other hand, a feeling of diminutiveness and need for protection may be conveyed by wearing higher heels, which make the step noticeably shorter.

Such knowledge of the human body has remained alive for thousands of years. Yet in the intervening time, in the field of development and behavior patterns, hardly any scientific facts worth teaching have been established.

APPENDIX

with a GLOSSARY *and* SELECTED REFERENCES

1. Rationale for Renaming Components of the Human Conceptus

Regional comparisons of human conceptuses, particularly during the early stages of ontogeny, have yielded many new findings that highlight differences between the behavior of the human germ and those of other animals. These new findings demand a neutral nomenclature so that the concepts of early human development are not prejudiced by the findings of experimental studies in animal embryos. For example, it is explained in Chapter 2 why the morula stage does not exist in human ontogeny and why the human embryo does not gastrulate.

The new terminology, which has been introduced in Chapter 2, is summarized in the schema on the following page—the suffix "-blast" is reserved for the **conceptus** as a whole and the suffix "-derm" for the **embryo**, which is a component of the conceptus. The prefixes "ecto-," "endo-," and "meso-" refer to the outside, inside, and middle layer, respectively, of the conceptus and also of the embryo. Further details with diagrams are given in Chapter 2 and the terms are defined in the Glossary

After implantation the conceptus consists initially of two components, outer and inner. The outer component that confronts the maternal tissue is termed the **ectoblast**; the ectoblast utilizes the nutrition released by dying maternal cells and grows rapidly in surface area and volume. The inner component is termed **endoblast**; the endoblast consists of a single pool of fluid (former blastocyst fluid) that is bordered by a dense cluster of cells at one side and elsewhere by flattened cells. In

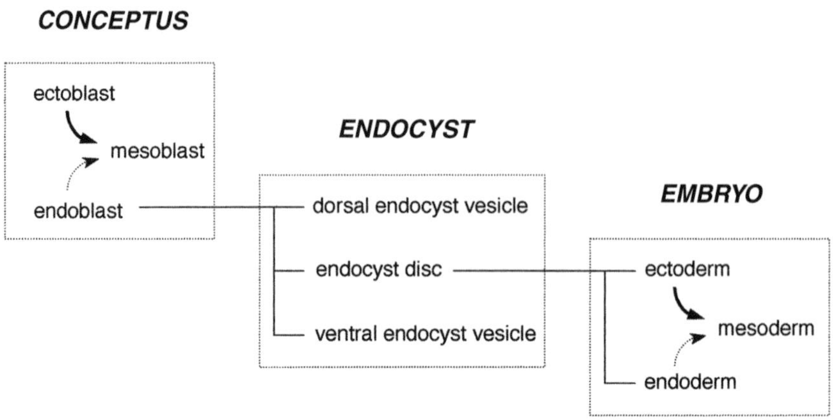

CONCEPTUS

Fig. A1. Schematic representation of the relations between conceptus, endocyst, and embryo.

contrast to the ectoblast, the endoblast does not enlarge so rapidly and so its growth lags behind that of the ectoblast. Thus, in about the 2nd week after fertilization, the ectoblast glides away from the endoblast and so creates an intermediate zone of cells and fluids in the conceptus. This newly formed tissue is the middle component of the conceptus and is termed the **mesoblast**; the mesoblast arises by a kind of shearing or **dehiscence** caused by differential growth.

Meanwhile, the volume of the endoblast is slowly increasing and a second pool of fluid (a sign of metabolism) has formed within it; this chamber is the anlage of the amniotic sac. The endoblast now consists of two pools of fluid and a curved disc of cells sandwiched between them. The collective name for the double-chambered endoblast with its covering mesoblast is the **endocyst**. The curved disc between the two pools of fluid is the **endocyst disc**.

The endocyst disc is the double-layered **anlage of the embryo**. On account of their positional relations to the subsequent embryo, the original chamber of fluid in the endoblast is called the **ventral endocyst vesicle** (anlage of yolk sac) and the second chamber of fluid is called the **dorsal endocyst vesicle** (anlage of amniotic sac).[24] The layer of cells in the endocyst disc that forms the "floor" of the dorsal endocyst vesicle is a tall, rapidly growing wedge-shaped epithelium that becomes the **ecto-derm** of the embryo. The shorter cells of the endocyst disc that "roof" the ventral endocyst vesicle become the **endoderm** of the embryo. In about the 3rd week after fertilization, the phenomenon of growth shearing that previously led to the development of the mesoblast of the whole conceptus repeats itself within the tissues of the endocyst disc: the faster growing ectoderm glides away from the endoderm creating an intermediate zone of cells and fluids termed the **mesoderm** of the embryo.

The repetition of the formation of an intermediate zone between a rapidly growing component and a less rapidly growing component is the earliest example of recapitulation in human ontogeny; it is also the first example of encaptic growth in human ontogeny. The chamber of fluid that arises in the mesoderm of the embryo is termed the **intra-embryonic coelom**; the chamber of fluid in the mesoblast of the whole conceptus is the **extra-embryonic coelom** (or chorionic sac, or preventral vesicle).

The main source of mesoblast in the conceptus is its most rapidly growing component, namely the ectoblast; we can envisage cells and fluids being squeezed from the ectoblast and being left behind as the latter grows away. Similarly, the main source of mesoderm in the embryo is its ectoderm; cells are squeezed from the ectoderm in the vicinity of the rapidly enlarging single brain bulge, the **expansion dome**, and also in the depressed region at the inferior end of the embryo, the **impansion pit**. In principle, there is no difference between the formation of the mesoblast of the conceptus and the mesoderm of the embryo. Furthermore, there are similarities between the formation of mesoderm and neural crest cells, so that, in the head region of the embryo, the inner tissue derived from ectoderm has long been described as **mesectoderm** or ectomesenchyme. In many regions of the embryo, there exists the potential for displacement of the limiting ectodermal cells into the underlying (inner) region on account of lateral pressure as the sheet of ectoderm expands against the resistance of the amnion. Just as the wet pit of a fruit can be expelled effortlessly and rapidly between the lips, so it is likely that the **polyingression** of ectoderm to form mesoderm is a consequence of the ectodermal cells being squeezed and is not an active process on the part of the ingressing cells. "Cell migration" is another example of a *deus ex machina* that obscures our understanding of the global nature of growth.

Some endodermal cells and fluids could also be squeezed into the mesoderm layer, particularly in the vicinity of the impansion pit; similarly, it is impossible to exclude a small contribution from the endoblast to the mesoblast of the conceptus. These possible contributions are indicated by the dotted lines in the above diagram.

2. A Simple Calendar of Human Development

1st week

Development from fertilization to the beginning of implantation. Special stage: the *one*-chambered conceptus (blastocyst).

2nd week

Complete implantation. Special stage: the *two*-chambered conceptus containing the endocyst disc (anlage of the embryo).

3rd week

Folding of the endocyst disc and development of the *embryo*. Special stage: the *three*-chambered conceptus with dorsal (amniotic), ventral (yolk sac), and preventral (chorionic) blastemal fluids.

4th week

Distinct head, neck, and trunk regions of embryo. Beginning of closure of ventral abdominal wall. Formation of large organ systems: brain, spinal cord, nerves, precartilaginous skeleton, musculature, and viscera (heart with atria and ventricles, liver with two lobes). Characteristic: development of metamerism with the formation of most pairs of somites (up to about thirty).

2nd month

Formation of umbilical cord. Early development of almost all definitive organs. Start of ossification but skeleton still mostly cartilaginous. First reflex movements of facial muscles.

3rd month

Start of fetal development, characterized by large cranium and small face, slender extremities.

4–10th lunar months: Late intrauterine development and birth.

Adapted from Blechschmidt (1974)

3. Relative Size of Embryo, Chorionic Sac, and Yolk Sac

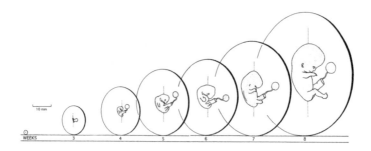

Fig. A2. The illustration shows the relative sizes of the chorionic sac, embryo, and yolk sac for Carnegie Stages 6, 10, 13, 16, 17, 20, and 23; Carnegie Stages are described in more detail in the following table (from O'Rahilly & Müller, 1987). *Reproduced with permission of the publisher.*

4. Carnegie Developmental Stages in Human Embryos

Carnegie Stage	Pairs of Somites	Length (mm)	Age (days)	Main Features
I		0.1–1.5		Fertilization: unicellular stage.
2		0.1–0.2	1.5–3	Blastomeric ovum with 2 to about 16 cells.
3		0.1–0.2	4	Free blastocyst.
4		0.1–0.2	5–6	Adplantation—blastocyst attaching to uterus.
5		0.1–0.2	7–12	Implantation—implanted (previllous stage).
5a		0.1	7–8	Solid ectoblast (trophoblast).
5b		0.1	9	Ectoblastic lacunae arise.
5c		0.15–0.2	11–12	Maternal blood enters ectoblastic lacunae.
6		0.2	13	Ectoblastic (chorionic) villi; impansion pit.
6a				Ectoblastic (chorionic) villi increase.
6b				Increase in area of expansion dome.
7		0.4	16	Axial process.
8		1.0–1.5	18	Impansion pit leads into axial (notochordal) canal; single expansion dome → two brain bulges.
9	1–3	1.5–2.5	20	Somites first appear; neural groove deepens between brain bulges.
10	4–12	2–3.5	22	Neural folds begin to fuse in neck region; two visceral arches form; optic sulcus arises.
11	13–20	2.5–4.5	24	Superior neuropore closes; optic vesicle forms.
12	21–29	3–5	26	Inferior neuropore closes; three visceral arches; upper limb buds appear. Length of embryo may decrease due to growth bending (flexion).
13	30–?	4–6	28	Four limb buds; lens placode.

Carnegie Stage	Pairs of Somites	Length (mm)	Age (days)	Main Features (continued)
14	5–7	32		Lens pit and optic cup; endolymphatic sac distinct; lung buds; S-shaped heart.
15	7–9	33		Lens vesicle; nasal pit; antitragus of auricle; flat metacarpus ("hand plate"); trunk relatively wider; cerebral hemispheres distinct.
16	8–11	37		Nasal pit faces ventrally; retinal pigment visible in intact embryo; lens still vesicular; auricle developing; metatarsus ("foot plate").
17	11–14	41		Head relatively larger; trunk extending (straightening); frontonasal groove distinct; auricle distinct; fingers.
18	13–17	44		Body more cuboidal; elbow region and toes appearing; eyelids forming; solid lens; tip of nose distinct; nipples appear; genital tubercles; ossification begins in clavicle.
19	16–18	47.5		Trunk elongating with growth extension of head–neck; ossification in mandible.
20	18–22	50.5		Upper limbs longer and flexed at elbows; cochlear duct has ½ turn; dorsiflexion of foot.
21	22–24	52		Fingers longer; hands pronated and feet inverted approaching median plane; cochlear duct ¾ turn.
22	23–28	54		Eyelids and external ear more developed; cochlear duct has one turn.
23	27–31	56.5		[end 2nd month] Head more rounded; limbs longer.

Modified from O'Rahilly & Müller (1987) and Hinrichsen (1990)

5. Variability in Normal Human Embryos

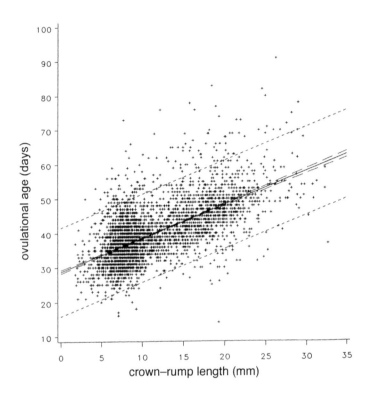

Fig. A3. This graph shows the variability between embryonic length (crown–rump length in mm) and ovulational age (in days; estimated by subtracting 14 days from the menstrual age) for 3,746 embryos. Note the large variation in length at any particular age. The line indicates the linear regression relationship: Carnegie Stage = 28.69 + 0.98 x length. The dotted lines indicate 95% intervals (from Shiota, Fischer, & Neubert, 1988). *Reproduced with permission of the publisher.*

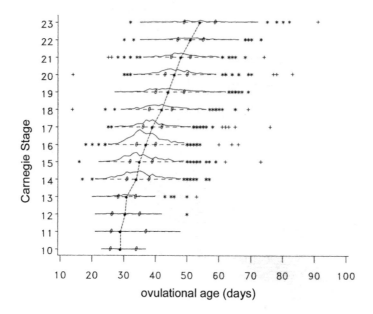

Fig. A4. This graph shows the great variability between ovulational age (estimated by subtracting 14 days from the menstrual age) and the Carnegie Stage based on morphological characteristics for 3,849 human embryos; • represents median value of age by Carnegie Stage; 0 represents lower and upper quartiles; * and + represent extreme values (from Shiota, Fischer, & Neubert, 1988). *Reproduced with permission of the publisher.*

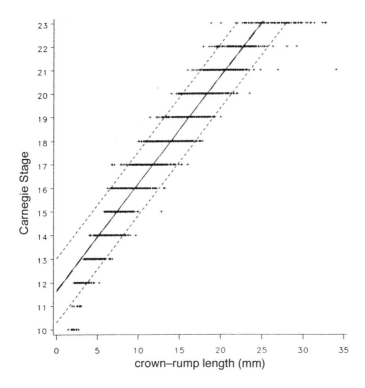

Fig. A5. This graph shows the relationship between embryonic length (crown–rump; in mm) and Carnegie Stage for 5,129 embryos (+). The line is the linear regression relationship: Carnegie Stage = 11.6 + 0.45 x length; the dotted lines are 95% intervals. Compared to the variability in the preceding two figures, this graph indicates that the length of the embryo has a more precise relation with the Carnegie Stage than the estimated age (from Shiota, Fischer, & Neubert, 1988). *Reproduced with permission of the publisher.*

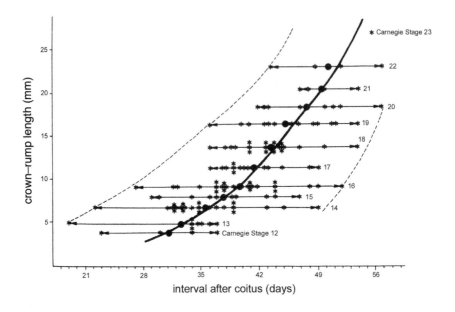

Fig. A6. This graph shows the distribution of embryonic length (crown–rump length in mm) for 135 single embryos recovered after a single isolated intercourse (interval after coitus; days). Each asterisk indicates an individual embryo and the solid circles indicate the mean age for each developmental stage (Carnegie Stage; indicated by numbers against horizontal lines). Dotted lines indicate minimum and maximum intervals. Note the amazing variability in age that may be associated with a single Carnegie Stage and the fact that this variability is greater for greater numbers of embryos (e.g., compare Carnegie Stages 22 and 23 with 14 and 16). (Diagram from Shiota, Fischer, & Neubert, 1988). *Reproduced with permission of the publisher.*

6. Fetal Growth

AGE		LENGTH	WEIGHT
(end of) week	(end of) lunar month	crown–rump (cm)	(gm)
9–12	3	5–8	10–45
13–16	4	12–13	60–200
17–20	5	17–19	250–450
21–24	6	21–23	500–820
25–28	7	24–27	900–1300
29–32	8	28–31	1400–2100
33–36	9	32–35	2200–2900
37–40	10	34–36	3000–3400

Adapted from Blechschmidt (1960) and Sadler (2000)

Glossary

acrosome: cap-like vesicle containing enzymes covering half of nucleus of spermatozoon.

adplantation: attachment of conceptus to surface of uterine mucosa, before implantation.

afferent: toward, e.g., the heart, the brain, etc.

allantois: a component of endoderm in the body stalk; a diverticulum of the yolk sac that develops eventually into the urachus (from Greek, like a sausage).

alveolus: a small sac, e.g., in the lungs.

amnion: the wall component of the dorsal endocyst vesicle, containing the amniotic fluid (from Greek, a lamb).

anabolite: a product of anabolism, the building-up of the body's substance.

anastomosis *(pl. anastomoses)*: a junction between two adjacent structures, e.g., blood vessels, nerves.

anlage (pl. anlagen): primordium or precursor; initial temporary state (German, a laying on).

appositional: referring to growth on a pre-existing surface, as opposed to interstitial growth.

arachnoidea: the arachnoid membrane consisting of a delicate layer around the brain and spinal cord resembling a spider's web; lies between pia mater and dura mater.

archetype: a presumed original pattern or model after which something might be made.

articulation sling: tissue connection between the anlagen of so-called antagonistic muscles, leading to the establishment of joint spaces under the apex of the sling.

ascensus: the particular positional development of the neural tube in a cranial (upwards) direction relative to the embryonic viscera; see *descensus.*

atrium: a chamber or sac (of the heart, etc.).

autolysis: self-digestion.

axial process: a projection or extension of the ectoderm into the inner tissue below the expansion dome of the endocyst disc.

axon: see neurite.

basicranium: the early cartilaginous base of the skull.

biodynamics: dynamic aspects of forces acting in ontogeny, with respect to the fluctuating mechanical equilibrium of metabolic processes in the developing organism.

biokinetics: the kinetic (spatiotemporal) aspects of development of the organism.

biomechanics: the mechanical features of the development of the organism.

blastemal: related to a sprout or a rudiment; see anlage.

blastocoele: fluid-filled space in the blastocyst; the anlage of the lumen of the yolk sac.

blastocyst: the one-chambered conceptus.

blastodisc: zone of the thick-walled part of the blastocyst; the inner cell mass.

blastomere: a daughter cell from the subdivision of the fertilized ovum.

blastomeric ovum: early multicellular stage (with many blastomeres) in human ontogeny (incorrectly called a morula).

body stalk: tissue bridge connecting endocyst and chorion; anlage of part of umbilical cord.

branchial: related to the gills (of fishes).

ca.: abbreviation for Latin *circa*, about.

calyx (pl. calyces): a cup-shaped space or region.

calvaria: the skull-cap.

canalization: the process of fluid movements giving rise to intercellular pathways or channels.

capacitation: phenomenon of exposure (over some hours) of spermatozoa to fluids of female genital tract; required for normal acrosomal reaction of fertilization.

capillary: very thin-walled blood vessel; beginning of all vessels in embryo (Latin, hairlike).

cartilage: translucent gristle containing cells known as chondrocytes.

catabolite: a product of catabolism, which is the breaking-down of the body's substance.

caudal: toward the tail; inferiorly.

cerebralization: brain development as the major partial process of total development.

choana (pl. choanae): a funnel-shaped space; the posterior nostrils.

chordae tendineae: "heart strings" made of collagen connecting valves to walls of ventricles.

chorion: wall of chorionic sac consisting of ectoblast (trophoblast) and mesoblast (extra-embryonic mesoderm); later forms the smooth chorion (chorion laevae) and the bushy chorion (chorion frondosum).

chorionic sac: fluid-filled space surrounding the endocyst as the extra-embryonic coelom.

choroid plexus: vascular gland-like formation, resembling the chorion, within the ventricles of the brain.

chromosomes: thread-like structures of high molecular weight observed during cell division.

cloaca: name given to chamber attached to hindgut, allantois, and mesonephric duct in human embryo; common excretory chamber in birds, most fishes, and monotremes.

coelom: early body sac with intra-embryonic and extra-embryonic components; also a transient space in umbilical cord.

collagen: main supportive protein of inner tissue (tendon, skin, cartilage, bone, etc.); converted to gelatin on boiling.

commissure: band of fibers joining right and left sides of brain or spinal cord.

conceptus: product of fertilization; generic term for totality of cells and fluids derived from the fertilized ovum; the embryo is one component of the conceptus.

conchae: scroll-like bones that project from the lateral wall into chamber of nose.

congenital: present at birth.

connecting stalk: connection between embryo and chorion incorporated into umbilical cord.

connective tissue: supporting tissue of the adult derived from inner tissue of embryo.

contusion field: metabolic zone where cells of inner tissue are pushed together, so becoming flatter and discoidal; field of precartilage cells (future chondrocytes).

cornified: converted to hard, horny material (keratin).

coronal: related to suture extending across skull between frontal and parietal bones.

corrosion field: metabolic region where epithelial cells die due to poor metabolic exchange.

cranial: toward the head; superior.

cranial nerve: a peripheral nerve connected to the brain containing sensory dendrites and/or motor neurites (see spinal nerve).

cuneiform: wedge-shaped.

cytoplasm: the matter of a cell outside the nucleus and within the boundary membrane.

decidua: part of endometrium reacting to implantation; later consisting of basal, capsular, and parietal regions.

dehiscence: an opening-up as occurs in mature ovarian follicle, fetal vagina, a wound, etc.

dendrite: extension of nerve cell with capacity to conduct stimuli toward cell body (soma).

densation field: a metabolic field where small cells of inner tissue crowd together as they lose water; a zone of precartilage; a premuscle mass.

-derm: suffix indicating a distinct layer; often "skin-like" (e.g., ectoderm, epidermis).

dermatome: component of a somite adjacent to ectoderm.

dermis: lower layer of the skin lying under the epidermis (also "true" skin or corium).

descensus: positional development of viscera in an inferior direction relative to brain.

detraction field: metabolic region where inner tissue is subject to compression and tensile stress in such a way that local dehydration occurs with deposition of intercellular solids; zone of ossification.

developmental dynamics: the kinetic and dynamic manifestations of differentiation.

developmental functions: cooperative performance of several organs during development.

developmental movements: constructive (forming) movements including submicroscopic material movements; the manifestations of spatially ordered metabolic movements.

diathelium: limiting tissue that separates two different fluids or media.

differentiation: the development of different parts of the body relative to each other.

dilation field: metabolic zone where the cells of inner tissue are easily extended and capable of lateral growth; all muscle cells arise in dilation fields.

diploid: having normal number (forty-six) of paired chromosomes in somatic cells.

distal: farther from the center, as opposed to proximal.

distusion field: metabolic zone of cells of inner tissue where (discoid) cells have a high osmotic pressure and start to swell, exerting a piston-like action on surrounding cells; field of chondrocyte hypertrophy.

diverticulum: a blind tubular sac leading away from a larger sac or tube.

dorsal: posteriorly; behind.

dorsal endocyst vesicle: a positional term for the anlage of the amniotic sac.

dura mater: outer, hard membrane covering spinal cord and brain.

ectoblast: external limiting tissue of the conceptus, also known as trophoblast.

ectoderm: initially the dorsal, later the external, limiting tissue of the endocyst disc and the embryo; an especially powerful layer of cells in early stages of ontogeny.

ectomeninx: the outer layer of the meninx or "skin" around the brain.

efferent: leading or conducting away from, e.g., the heart, the brain, etc.

embryo: developing human organism that arises from the endocyst disc of the conceptus from about 4–8th week after fertilization.

embryology: the science of the embryo and its development.

encapsis: subdivision of a whole into parts that have a formal similarity with the whole.

endoblast (entoblast): inner limiting tissue of conceptus, consisting initially of inner cell mass, blastocyst fluid, and its bordering flat cells; later comprising dorsal endocyst vesicle (anlage of amniotic sac) and ventral endocyst vesicle (anlage of yolk sac) with endocyst disc in between.

endocrine glands: glands with ruptured excretory ducts; sometimes called incretory glands.

endocyst: the two-chambered endoblast (dorsal and ventral endocyst vesicles, or amniotic sac and yolk sac, respectively) together with covering mesoblast.

endocyst disc: the human germ disc in the endocyst; the anlage of the embryo between the dorsal and ventral endocyst vesicles.

endoderm (entoderm): initially the ventral, less powerful limiting tissue of the endocyst disc.

endomeninx: the inner layer of the meninx or "skin" around the brain.

enzyme: a protein catalyst.

epidermis: outer layer of skin; above the dermis.

epithelium: see limiting tissue.

evagination: a turning-inside-out; protrusion of a part of an organ.

evolution: the history of development as distinct from ontogeny (individual development).

excretory apparatus: the urinary organs.

exocrine glands: glands with excretory ducts.

expansion dome: dorsally arching part of endocyst disc; as opposed to the impansion pit.

extension: the act of straightening a limb or structure.

fascicle: a little bundle of collagen fibers, muscle fibers, nerve fibers, etc.

fertilization: the union of the ovum and the sperm to produce the conceptus.

fetus: developing human between start of 3rd month of gestation and birth.

fixation: chemical or physical preservation of tissues for histology (by a fixative agent).

flexion: the act of bending a limb or structure.

foramen (pl. foramina): a passage or opening.

forming function: organic performance that manifests itself as a formation, i.e., as a morphologically discernible structure. All anlagen including fluids in body sacs have forming functions (German *Gestaltungsfunktion*).

frontal: at right angles to the sagittal plane; in a coronal plane.

functional development: the development of performances in ontogeny; all organic differentiation is a development of position, form, structure, and function.

functionalism: teleological doctrine holding that form and constitution are determined primarily by presumed function.

fundus: part or base of an organ remote from its opening, e.g., vitreal surface of retina of eye.

ganglion (pl. ganglia): a cluster of nerve cells outside the brain and spinal cord.

gastrulation: process whereby the spherical embryo of amphibians and certain fishes becomes two-layered by invagination of part of the wall; the same process does not occur in humans.

genes: hereditary factors of cell nucleus that are partial components of all metabolic fields; the nuclear sites for the application of external, differentiating forces.

genome: totality of genetic material in a cell; diploid set of chromosomes in a somatic cell.

germ: the first anlage of an organism; conceptus.

gestation: the period of time from conception to birth.

glia: supportive tissue for neurons (Greek, glue).

glottis: the sound-producing part of larynx, consisting of vocal folds and intervening space or rima glottidis (from Greek, the back of the tongue).

glycocalyx: complex of carbohydrate and protein covering the cell boundary membrane.

gonads: the embryonic sex glands; a generic term for ovaries and testes.

ground substance: fluid or material occupying the intercellular spaces in inner tissue.

growth functions: performances of growing organs due to submicroscopic movements of metabolic materials.

gubernaculum: rudder or helm; guidance structure, as in g. dentis, g. nasi, g. testis.

gyrus (pl. gyri): a convolution of the cerebral cortex demarcated by sulci or fissures.

Hensen's knot (node): clump of cells seen after chemical fixation of some embryos at cranial end of primitive streak at site of transition from expansion dome to impansion pit; described in guinea pig embryo by V. Henson, German physiologist, in 1876.

histology: study by microscopy of artificially colored tissues and organs.

homunculus: a diminutive human or human essence.

hyoid: related to the U-shaped bone at the base of the tongue.

impansion pit: lower, dorsally concave part of endocyst disc; in contrast to expansion dome.

implantation: establishment of conceptus within uterine mucosa following adplantation.

incipient: beginning; coming into existence.

induction: influencing differentiating processes (at a distance) by means of chemical substances.

inner cell mass: the blastomeres that constitute the more cellular part of the endoblast.

inner tissue: cells and intercellular substances enclosed by limiting tissue, e.g., connective tissue.

inner tissue of the conceptus: mesoblast.

inner tissue of the embryo: mesenchyme.

inner tissue of the endocyst disc: mesoderm.

intercellular substance: material located in interstices between cells; see ground substance.

intercostal: between the ribs.

interstice: a space between cells (also called interstitium).

invagination: an ingrowth or ensheathing; an infolding of a portion of the wall of an organ.

labyrinth: a system of tortuous membranous tubes and spaces derived from the otocyst (membranous labyrinth) or the spaces in the temporal bone that contain the membranes (bony labyrinth); the inner ear.

lamina: thin flat layer or membrane.

lateral: to the side.

leptomeninx: soft inner part of meninx (pia mater + arachnoid) relative to pachymeninx.

limiting (boundary) tissue: the intervening layer of usually wedge-shaped cells between a fluid on one side and inner tissue on the other; diathelium.

lingual: of the tongue (Latin *lingua*).

lumen: space within a tube, or vesicle, usually filled with fluid.

marginal mesoblast: the inner tissue at the margin of the endocyst disc.

mastication: chewing.

meatus: a passage or opening.

medial: toward the midline (median) plane of the body.

median: of the mid-plane dividing the body into right and left halves.

meninx (pl. meninges): one of the three membranes ("skins") around the brain and spinal cord; composed of pachymeninx (dense layer) and leptomeninx (soft layer) in embryo.

mesectoderm: inner tissue of embryo derived from surface ectoderm of head region.

mesenchyme: inner tissue of the embryo derived from mesoderm and/or mesectoderm; consisting of cells, intercellular fibers, and intercellular fluids of inner tissue.

mesoblast: inner tissue of conceptus (also called extra-embryonic mesoderm); exists later as covering mesoblast (outer layer of endocyst) and lining mesoblast (inner layer of ectoblast or chorion); also see marginal mesoblast.

mesoderm: inner tissue of the endocyst disc.

mesonephros: embryonic excretory (urinary) apparatus; see metanephros.

metabolic field: a region of metabolism, determined by its morphological and biodynamic properties, containing spatially ordered metabolic movements.

metabolic fields, biodynamic: metabolic fields with respect to their biodynamic significance.

metabolic movements: submicroscopic material movements in a morphologically definable metabolic field.

metameric: segmental; following one another in a step-like fashion.

metanephros: anlage of the adult or definitive kidney (see mesonephros).

mitosis: cell division where each daughter cell contains the same number of chromosomes in the nucleus as the parent cell; the state of chromosomes becoming thread-like.

morphogenesis: formal development; the forming of structures.

morula: term applied by E. Haeckel in 1876 to a free-swimming stage of a coral embryo (Latin, a little mulberry); term does not apply to human development where the blastomeres are surrounded by a zona pellucida and the anlage of the blastocoele arises with the very first subdivisions of the ovum (see blastomeric ovum).

mucosa: the limiting tissue and its stroma lining the viscera, e.g., uterus, intestines, etc.

myotome: component of a somite; the early musculature of the back.

nascent: just being born; incipient; beginning (see anlage).

neural crest cells: cells deposited from the basal (abluminal) side of the ectodermal crests at the lateral margins of the neural groove;

neural crest cells give rise to ganglia, connective tissues, etc. (see mesectoderm; polyingression); the cells do not "migrate" through the embryo but are rather anchored as other cellular ensembles grow past them.

neural groove: early, longitudinal groove arising in ectoderm of embryo as the anlage of brain and spinal cord.

neural tube: early tube-like anlage of brain and spinal cord after closure of neural groove.

neurite: extension of nerve cell conducting signals away from cell body (to muscles, etc.).

neurocoele: fluid-filled lumen of neural tube.

neuron: nerve cell with cell body (soma), neurite, and dendrites.

neuropore: part of neural tube that remains open to amniotic sac for a short time.

nidation: a nesting of the blastocyst in the endometrium.

notochord: the column of cells arising from the axial process.

occiput: the back of the head.

omphalo-enteric: referring to the vitelline stalk between the yolk sac and the midgut.

ontogeny: development of an individual.

ontological: resulting from the one fertilized ovum.

ontology: the investigation of the nature of being.

osmotic pressure: pressure due to concentration difference of solutes across a membrane.

ossification: process of bone formation.

otic: of the ear.

otocyst: the early vesicular stage of the inner ear that arises from the otic placode.

ovary: the female genital gland producing the ovum.

ovum: female germ cell; also generic term for female egg or conceptus for about 2–3 weeks after fertilization prior to development of embryo, in which case it is described as the fertilized ovum, the blastomeric ovum, the three-chambered ovum, etc.

pachymeninx: dura mater (see leptomeninx).

paleontology: the investigation of life-forms of former geological periods.

parathelial: of loosening inner tissue where limiting tissue is extending to form a gland, etc., in a suction field.

parietal: related to the lining of the wall of a sac; as opposed to visceral.

parthenogenesis: reproduction from a female egg without fertilization by male.

performances: achievements or functions emerging in any particular organ at any particular time in ontogeny.

perichondrium: layer or "skin" of fibrous inner tissue on the surface of cartilage.

periosteum: layer or "skin" of fibrous inner tissue on surface of bone.

peristaltic: of the wave-like narrowing of intestinal lumen passing from superior to inferior.

peritoneum: serous membrane reflected over the viscera and lining abdominopelvic (peritoneal) sac.

permeable: of a capacity of submicroscopic particles, molecules, etc., to move unimpeded.

petrous: hard as a rock.

pharyngeal arches: see visceral arches.

phenogenesis: differentiation; the process of becoming visible.

phenotype: the physical appearance or make-up of an individual.

phylogenesis: the presumed history of evolution of races, species, tribes.

pia mater: richly vascularized inner tissue on outer surface of brain and spinal cord.

pineal gland: a small gland shaped like a pine cone.

placenta: part of conceptus forming interface with mother's uterus (from Latin, a flat cake).

placode: local thickening of limiting tissue (often ectoderm) as the anlage of an organ.

plexus: a braiding of vessels or nerves.

polyingression: the formation of embryonic inner tissue directly from ectodermal cells that detach from the surface at multiple sites (see mesectoderm).

portal: an entry way, e.g., the superior and inferior intestinal portals, the right and left coelomic portals.

precartilage: closely packed small cells in a densation field; cell size decreases due to gradual loss of water with cell division; cells become discoidal chondrocytes in a contusion field.

preventral vesicle: the chorionic sac or extra-embryonic coelom.

primitive streak: midline depressed relief at inferior end of S-shaped embryonic disc at the impansion pit ("primitive" indicates the relative time of development, not its nature!).

process: (i) a series of events; (ii) an anatomical (structural) projection on a bone, from an epithelium, etc.

procollagen: precursor of tropocollagen, which polymerizes intercellularly as a collagen fibril.

pronation: a rotation of the forearm taking the palm away from the face (toward the ground) with crossing of forearm bones; as opposed to supination.

proximal: nearer to the trunk (or other point of reference); as opposed to distal.

psychic: of the human mind; metaphysical; as opposed to somatic or physical.

Rathke's pouch: ectodermal diverticulum giving rise to adenohypophysis (part of so-called pituitary gland); described by M.H. Rathke, German anatomist, in 1861.

renal: related to the kidney (Latin *ren*, a kidney).

retension field: metabolic field of inner tissue in which cells are extended and transversely compressed, thereby offering a resistance to further stretch; all dense connective tissue arises in retension fields.

retinaculum: a band of tissue that holds other structures (e.g., tendons) in place.

retroperitoneal: of organs located behind the peritoneum at the posterior wall of the abdominal sac and deep to the peritoneum of the pelvis, e.g., kidneys, suprarenal glands, pancreas, rectum, bladder, etc.

rudiment: an anlage; a part that is undeveloped.

sac: a fluid-filled chamber sometimes referred to as a "cavity."

sagittal: parallel to the median plane (containing the sagittal suture of the skull).

sclerotome: the ventromedial portion of a somite that contributes to the formation of the axial skeleton.

septum (pl. septa): wall dividing partially or completely two structures, spaces, or regions; e.g., the transverse septum of the embryo near the junction of the foregut and the remainder of the yolk sac.

serosa: lining of the body sac.

sesamoid bone: bone that arises within a tendon (in the shape of a sesame seed).

skeletonization: the process of forming the skeleton.

soma: body of a cell, etc.

somatic: of the body; of the body wall.

somites: small rounded, block-like organs of body wall in the back region of embryo; may contain a transient sac (somitocoele).

somitocoele: transient fluid-filled lumen of a somite.

spinal ganglion: cluster of nerve cells lateral to the spinal cord; also known as a dorsal root ganglion.

spinal nerve: one of the metameric peripheral nerves connected to the spinal cord containing sensory dendrites and motor neurites.

strain: spatial deformation produced by stress.

stress: biodynamically, force(s) acting in metabolic field(s).

stroma: inner tissue, especially underlying a limiting tissue.

suction field: metabolic field in which an epithelial sheet is subject to suction forces by surrounding growth movements, thereby allowing the epithelial cells to extend easily by surface growth into adjacent inner tissue. All glands (e.g., lung, liver, thyroid, sweat, etc.) arise in suction fields.

sulcus (pl. sulci): groove or furrow.

supination: rotation of forearm bringing palm to the face (or sky) and forearm bones parallel.

suprarenal: lying near superior pole of kidney.

synovial: like egg white.

tectogenesis: development of the internal structure of an organ; as opposed to development of its external form (morphogenesis).

teleological: of the study of evidence for a presumed design or purpose in nature; doctrine of final causes (see functionalism).

teleonomy: laws that govern a presumed design or purpose in nature.

tensile stress: stress that pulls, as distinct from compression.

topogenesis: positional development of an organ, structure, etc.

trabecula (pl. trabeculae): a little beam of connective tissue, bone, cardiac muscle (trabeculae carneae); a spicule of bone.

trajectorial: of a series of curves or surfaces that intersect another series of curves or surfaces according to a given rule, e.g., at a constant angle.

trophoblast: ectoblast; the thickest outer part of conceptus that transfers nutrition (*trophe*) to endoblast; later composed of two histological cell types: cytotrophoblast and syncytiotrophoblast.

umbilicus: the former perimeter of the embryonic disc where it becomes the amniotic ectoderm; the umbilicus becomes more ventrally located as the embryo grows dorsally; the navel.

urachus: urine-containing intra-abdominal part of fetal allantois and urogenital sinus; becomes the fibrous cord of middle umbilical ligament in adult (Greek *ourachos*).

uterine tube: tube carrying the ovum to uterus (womb); Fallopian tube.

vacuole: a space in the cell cytoplasm (or in inner tissue) filled with fluid.

vacuolization: process of congestion of fluids in interstices of inner tissue of conceptus, embryo, etc.

vascular: related to blood vessels.

vascularization: process of blood vessel formation as a consequence of a metabolic gradient.

ventral: toward the belly or front of the body.

ventral endocyst vesicle: anlage of yolk sac in the two-chambered endocyst.

ventricle: a little chamber; right and left ventricles of heart; chambers in brain.

vesicle: bubble or cyst containing fluid.

villus (pl. villi): a short, filamentous projection on a surface, e.g., intestinal villi, chorionic villi.

visceral: related to viscera or to that part of the lining of the body sac covering the viscera.

visceral arches: externally visible arch-like parts of flexion folds in face and ventral neck region of embryo; often described as pharyngeal arches but extending beyond the pharynx (throat).

vitelline: of the yolk of an egg; related to the yolk sac of a conceptus.

wedge epithelium: limiting tissue with diverging or converging wedge-shaped cells.

yolk sac: conventional name for the sac derived from the ventral entocyst vesicle; the fluid is watery, quite unlike the yolk of a chicken egg.

zona pellucida: thick glycocalyx around oocyte, ovum, and blastomeric ovum.

Selected References

Selected works by Erich Blechschmidt [25]

Blechschmidt, E. (1934) Die Architektur des Fersenpolsters. *Gegenbaurs Morphologisches Jahrbuch* 73: 20–68; translated and republished as: Blechschmidt, E. (1982) The structure of the calcaneal padding. *Foot & Ankle* 2: 260–283.

Blechschmidt, E. (1940) Über die Grundlagen zu enigen neuen Problemen der Entwicklunsgeschichte (Gesetzmäßigkeiten in der menschlichen Frühentwicklung). *Berichte der Physikalisch Medizinische Gesellschaft zu Würzburg*, N.F. 64: 11–60.

Blechschmidt, E. (1944) Über Massenbewegungen als Ursache der Körpergestaltung (Kausale Morphologie des Zahnschmelzes). *Nachrichten der Akademie der Wissenschaften in Göttingen, Mathematisch-Physikalische Klasse*, 7: 91–120.

Blechschmidt, E. (1947) Über das Formbildungsvermögen des Menschlichen Körpers (Die Gestaltungskraft des Epithels). *Abhandlungen der Akademie der Wissenschaften in Göttingen, Mathematisch-Physikalische Klasse*, Folge 3, 22: 1–44.

Blechschmidt, E. (1948) *Mechanische Genwirkungen*. Göttingen: Musterschmidt.

Blechschmidt, E. (1951a) Die frühembryonale Lageentwicklung der Gliedmaßen. (Entwicklung der Extremitäten beim Menschen. Teil I.) *Zeitschrift für Anatomie und Entwicklungsgeschichte* 115: 529–540.

Blechschmidt, E. (1951b) Die frühembryonale Formentwicklung der Gliedmaßen. (Entwicklung der Extremitäten beim Menschen. Teil II.) *Zeitschrift für Anatomie und Entwicklungsgeschichte* 115: 597–616.

Blechschmidt, E. (1951c) Die frühembryonale Strukturentwicklung der Gliedmaßen. (Entwicklung der Extremitäten beim Menschen. Teil III.) *Zeitschrift für Anatomie und Entwicklungsgeschichte* 115: 617–657.

Blechschmidt, E. (1953) Die Entwicklung der Zahnkeime beim Menschen. Zum Studium der Entwicklungsdynamik der menschlichen Embryonen. *Acta anatomica* 17: 207–239.

Blechschmidt, E. (1955a) Die Entwicklungsbewegungen der Zahnleiste. Funktionelle Faktoren bein der Frühentwicklung des menschlichen Kauapparats. *Roux' Archiv für Entwicklungsmechanik* 147: 474–488.

Blechschmidt, E. (1955b) Regional-vergleichende Untersuchungen von Differenzierungsbewegungen. Die Entwicklungsbewegungen im Differenzierungsgebiet von Osteoblasten und ihre funktionelle Bedeutung für den Zahndurchbruch. *Roux' Archiv für Entwicklungsmechanik* 148: 72–91.

Blechschmidt, E. (1955c) Embryologische Untersuchungen unter funktionellen Gesichtspunkten. *Acta anatomica* 24: 339–392.

Blechschmidt, E. (1956a) Entwicklungsfunktionelle Untersuchungen am Bewegungsapparat (Koordination von Entwicklungsbewegungen, Somatogenese). *Acta anatomica* 27: 62–88.

Blechschmidt, E. (1956b) Entwicklungsfunktionelle Untersuchungen am embryonalen Eigeweidesystem. Bauprinzipien der Eingeweide,

Beobachtungen zur Frage der funktionellen Bedeutung des Keilepithels und der ventrikulären Mitosen. *Morphologisches Jahrbuch* 96: 393–416.

Blechschmidt, E. (1960) *The Stages of Human Development before Birth. An Introduction to Human Embryology.* Basel: Karger.

Blechschmidt, E. (1963) *The Human Embryo. Documentations on Kinetic Anatomy.* Stuttgart: Schattauer.

Blechschmidt, E. (1966) Die Sprache der Hände. *Die Grünenthal Waage* 5: 12–24.

Blechschmidt, E., Daikoku, S. (1966) Die regionale Verschiedenheit embryonaler Dendriten und Neuriten (Elektronenmikroskopische Untersuchung). *Acta anatomica* 65: 30–57.

Blechschmidt, E. (1967a) Die Entwicklungsbewegungen der menschlichen Augenblase. *Ophthalmologica* 153: 291–308.

Blechschmidt, E. (1967b) Die Entwicklungsbewegungen der menschlichen Retina zur Zeit der Irisentstehung. Die Entstehung des Ganglion opticum als Beispiel einer submikroskopisch untersuchbaren Entstehung einer Cytoarchitectonik. *Ophthalmologica* 154: 531–550.

Blechschmidt, E. (1968a) Die Stoffwechselfelder des menschlichen Eis (Unser heutige Auffassung von der menschlichen Frühentwicklung). *Zeitschrift für Geburtshilfe und Gynaecologie* 168: 143–155.

Blechschmidt, E. (1968b) *Vom Ei zum Embryo.* Stuttgart: Deutsche Verlagsanstalt; 7th Ed., Stein-am-Rhein: Christiana-Verlag; translated and republished as: Blechschmidt, E. (1977) *The Beginnings of Human Life.* New York: Springer-Verlag.

Blechschmidt, E. (1969a) The early stages of human limb development. In: Swinyard, C.A. (Ed.) *Limb Development and Deformity: Problems of Evaluation and Rehabilitation.* Springfield: C.C. Thomas, pp. 24–56.

Blechschmidt, E. (1969b) Differenzierungen im kinetischen Feld (Enstehungsbedingungen der Metamerie). *Acta anatomica* 73: 351–371.

Blechschmidt, E. (1970) Die Entwicklungsdynamik der Visceralbögen (Funktionelle Differenzierungen). *Archiv für Hals-, Nasen- und Ohrenheilkunde (HNO)* 18: 263–271.

Blechschmidt, E. (1973) *Die pränatalen Organsysteme des Menschen.* Stuttgart: Hipprokrates.

Blechschmidt, E. (1974) *Humanembryologie. Prinzipien und Grundbegriffe.* Stuttgart: Hipprokrates.

Blechschmidt, E. (1977) The programming of afferent and efferent nervous fibers in man. *Archiv für Psychiatrie und Nervenkrankheit* 224: 259–272.

Blechschmidt, E. (1978) Der Systemcharakter der Zelle. *Scheidewege* 8: 527–534.

Blechschmidt, E., Gasser, R.F. (1978) *Biokinetics and Biodynamics of Human Differentiation. Principles and Applications.* Springfield: C.C. Thomas.

Other References

Arey, L.B. (1946) *Developmental Anatomy. A Textbook and Laboratory Manual of Embryology.* 5th ed. Philadelphia: Saunders.

Bray, D. (1984) Axonal growth in response to experimentally applied mechanical tension. *Developmental Biology* 102: 379–389.

Carey, E.J. (1920) Studies in the dynamics of histogenesis. Growth motive force as a dynamic stimulus to the genesis of muscular and skeletal tissues. *Anatomical Record* 19: 199–235.

Carey, E.J. (1921) Studies in the dynamics of histogenesis. IV. Tension of differential growth as a stimulus to myogenesis in the limb. V. Compression between the accelerated growth centers of the segmental skeleton as a stimulus to joint formation. VI. Resistances to skeletal growth as stimuli to chondrogenesis and osteogenesis. *American Journal of Anatomy* 29: 93–115.

Carey, E.J. (1922) Direct observations on the transformation of the mesenchyme in the thigh of the pig embryo (*Sus scrofa*), with especial reference to the genesis of the thigh muscles, of the knee- and hip-joints, and of the primary bone of the femur. *Journal of Morphology* 37: 1–77.

Freeman, B. (2003) The migration of germ cells in the embryos of mice and men is a myth. *Reproduction* 125: 635–643.

Gasser, R.F. (1979) Evidence that sclerotomal cells do not migrate medially during normal embryonic development of the rat. *American Journal of Anatomy* 154: 509–524.

Hinrichsen, K.V. (1990) *Humanembryologie. Lehrbuch und Atlas der vorgeburtlichen Entwicklung des Menschen*. Berlin: Springer.

Hinrichsen, K.V. (1992) In memoriam des Anatomen und Embryologen Erich Blechschmidt (1904–1992). *Annals of Anatomy* 174: 479–484.

O'Rahilly, R., Müller, F. (1987) *Developmental Stages in Human Embryos*. Washington: Carnegie Institution Publication 637.

Sadler, T.W. (2000) *Langman's Medical Embryology*. Philadelphia: Lippincott Williams & Wilkins.

Shiota, K., Fischer, B., Neubert, D. (1988) Variability of development in the human embryo. In: *Non-Human Primates—Developmental Biology and Toxicology*. Eds. Neubert, D., Merker, H-J., Hendrickx, A.G. Wien: Ueberreuter Wissenschaft, pp. 191–203, 240.

Sobotta, J. and Becker, H. (1962) *Atlas der Anatomie des Menschen*. Berlin: Urban & Schwarzenberg.

References with a different viewpoint

Eigen, M., Winkler, R. (1993) *Laws of the Game: How the Principles of Nature Govern Chance*. Princeton University Press.

Lorenz, K. (1977) *Behind the Mirror: A Search for a Natural History of Human Knowledge*. New York: Harcourt Brace Jovanovich.

Monod, J. (1971) *Chance and Necessity: An Essay on the Natural Philosophy of Modern Biology*. New York: Knopf.

NOTES

1. Development of an individual—words, or related terms, in bold are defined briefly in a Glossary in the Appendix.
2. Vesalius was born in Brussels; his family came from Wesel on the Rhine. At one time he was professor of surgery and anatomy in Padua, and then the royal physician to the Roman emperor Charles V, and later to Phillip II of Spain. The illustrations for the well-known woodcuts of his anatomical atlas may have been done by a pupil of Titian.
3. Parthenogenesis also occurs normally in some lizards and snakes, bees, aphids and other insects, and in domestic turkeys.
4. The phenomenon of parthenogenesis mentioned previously shows that significant differentiations may occur in the absence of male genes.
5. The Blechschmidt Collection and Museum of Human Embryos.
6. Carnegie Developmental Staging is a standardized method for classifying the developmental features of a human embryo and estimating its approximate age, according to morphological criteria that include the length of the embryo from crown to rump. A table describing Carnegie Stages is presented in the Appendix.
7. The basis for the new terminology of ectoblast, endoblast, mesoblast, etc., is presented in the Appendix.
8. The term "cavity" is avoided for this and other fluid-filled spaces or **sacs** in the conceptus.
9. The term *process* here refers to a structural (anatomical) projection and not an event.
10. The region of the impansion pit is described as **Hensen's knot** in animal embryos.

11. Cell division can occur in embryos by **mitosis** and also by amitosis (i.e., without the appearance of chromosomal threads).

12. In the Appendix, tables and graphs are given showing the ages of embryos with respect to the number of somites, crown–rump length, and other features. It can be seen from these figures that there may be considerable variability in the age of the embryo with respect to a given stage of development.

13. With the exception of the fields described in the pig embryo by the American anatomist Eben J. Carey (1899–1947), who analyzed differential growth and biomechanical zones in the development of musculoskeletal and visceral organs. Carey carried out an intensive investigation of a closely spaced series of pig embryos and emphasized the concept of regional comparison of ontological organs (see References) .

14. The spelling "retension" is used to avoid the associations of the word "retention."

15. The dynamics of the detraction field in the formation of bone were described before the concept of "distraction osteogenesis" (due to the Russian surgeon and scientist Gavriel Ilizarov) became common in the clinical literature of reconstructive surgery.

16. Cytoplasmic extensions of cell bodies.

17. The relation of spiral waves of endodermal mitosis to the development of gut muscles in the pig embryo was established by Eben J. Carey (see references).

18. Illies, J. *Zoologeleien*. Herderbücherei, Vol. 502, 3rd Ed., 1976. The original poem is: *Um diesen Menschen zu erschaffen,/da brauchte die Natur den Affen/—so hört man Brägengrütze sagen—/nur schnell vom Urwaldbaum zu jagen,/denn auf die Steppe ausgetrieben,/ist ihm nichts anderes geblieben.//Es wuchsen ihm seit jener Stunde/Bananen nicht mehr vor dem Munde;/er mußte, wollt' er weiter leben,/sich auf die Hinterbeine heben/und machte so die Hände frei/für Obst und sonst noch allerlei.//Die Hände greifen nach der Birne/und so entwickeln im Gehirne/(um alle Tricks gut zu behalten)/von Jahr zu Jahr sich neue Falten,/bis daß der Schädel heftig quillt/und sich mit tausend Gramm anfüllt.*

19. Latin *concipere*, to hold together; *percipere*, to receive.
20. Latin *referre*, to carry back; *sumere*, to take up; *reducere*, to bring back.
21. Latin *manus*, hand.
22. Latin *factum*, a thing done.
23. Latin *finis*, a boundary.
24. The chamber of fluid that arose previously in the mesoblast is therefore described positionally as the **preventral vesicle.**
25. See Hinrichsen (1992) under *Other references* for a complete list of Blechschmidt's publications.

INDEX

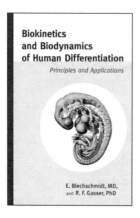
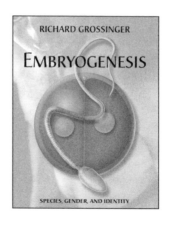

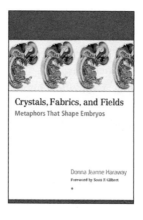

About North Atlantic Books

North Atlantic Books (NAB) is a 501(c)(3) nonprofit publisher committed to a bold exploration of the relationships between mind, body, spirit, culture, and nature. Founded in 1974, NAB aims to nurture a holistic view of the arts, sciences, humanities, and healing. To make a donation or to learn more about our books, authors, events, and newsletter, please visit www.northatlanticbooks.com.